CAMBRIDGE LIBRARY COLLECTION

Books of enduring scholarly value

Monographs of the Palaeontographical Society

The Palaeontographical Society was established in 1847, and is the oldest Society devoted to study of palaeontology worldwide. Its primary role is to promote the description and illustration of the British fossil flora and fauna, via publication of an authoritative monograph series. These monographs cover a wide range of taxonomic groups, from microfossils, trilobites and ammonites through to Coal Measure plants, mammals and reptiles, and from all ages from Cambrian to Pleistocene. They form a benchmark for understanding the past life of the British Isles and many include the original descriptions of numerous key species. The first monograph (on the Crag Mollusca) was published in March 1848 and the Society still continues this work today. Notable authors in the series include Charles Darwin (fossil barnacles) and Richard Owen (dinosaurs and other extinct reptiles). Beginning in 2014, the Cambridge Library Collection and the Society are collaborating to reissue the earlier publications, focusing on monographs completed between 1848 and 1918.

A Monograph on the British Fossil Echinodermata from the Cretaceous Formations

The magnificent monograph of the British Cretaceous echinoids (sea urchins) by Thomas Wright (1809–84) was to be followed by a similar work on the asteroids (starfishes). Sadly, Wright died in the early stages of this project, leaving only notes and some plates, but it was taken up by Walter Percy Sladen (1849–1900), who completed two parts before ill health interrupted his previous productivity. William Kingdon Spencer (1878–1955) became the third authority to be involved, finishing the work in a further three parts and also expanding the scope to include the ophiuroids (brittle stars). All three contributors were amateurs, variously a surgeon (Wright), independently wealthy (Sladen) and H.M. Inspector of Schools in Suffolk (Spencer). Originally published in five parts between 1891 and 1908, Sladen and Spencer's text, embellished by many fine plates, is a monument to two of the rarer, yet most attractive, groups of Cretaceous fossils.

Cambridge University Press has long been a pioneer in the reissuing of out-of-print titles from its own backlist, producing digital reprints of books that are still sought after by scholars and students but could not be reprinted economically using traditional technology. The Cambridge Library Collection extends this activity to a wider range of books which are still of importance to researchers and professionals, either for the source material they contain, or as landmarks in the history of their academic discipline.

Drawing from the world-renowned collections in the Cambridge University Library and other partner libraries, and guided by the advice of experts in each subject area, Cambridge University Press is using state-of-the-art scanning machines in its own Printing House to capture the content of each book selected for inclusion. The files are processed to give a consistently clear, crisp image, and the books finished to the high quality standard for which the Press is recognised around the world. The latest print-on-demand technology ensures that the books will remain available indefinitely, and that orders for single or multiple copies can quickly be supplied.

The Cambridge Library Collection brings back to life books of enduring scholarly value (including out-of-copyright works originally issued by other publishers) across a wide range of disciplines in the humanities and social sciences and in science and technology.

A Monograph on the
British Fossil Echinodermata
from the Cretaceous Formations

The Asteroidea and Ophiuroidea, Based on Plates by Thomas Wright

W. Percy Sladen
W.K. Spencer

CAMBRIDGE
UNIVERSITY PRESS

CAMBRIDGE
UNIVERSITY PRESS

University Printing House, Cambridge, CB2 8BS, United Kingdom

Cambridge University Press is part of the University of Cambridge.

It furthers the University's mission by disseminating knowledge in the pursuit of
education, learning and research at the highest international levels of excellence.

www.cambridge.org
Information on this title: www.cambridge.org/9781108081191

THE

PALÆONTOGRAPHICAL SOCIETY.

INSTITUTED MDCCCXLVII.

LONDON

MDCCCXCI—MDCCCCVIII.

MONOGRAPH OF THE BRITISH FOSSIL ECHINODERMATA FROM THE CRETACEOUS FORMATIONS.

VOL. II.—ASTEROIDEA AND OPHIUROIDEA.

Mr. Sladen is the author of pages 1—66, Plates I—XVI, while Mr. Spencer is the author of the remainder of the Volume. Mr. Spencer desires to express his indebtedness to Dr. F. A. Bather for much help and advice in his share of the work.

ORDER OF BINDING AND DATES OF PUBLICATION.

PAGES	PLATES	ISSUED IN VOL. FOR YEAR	PUBLISHED
Title-page	—	1908	December, 1908
1—28	I—VIII	1890	April, 1891
29—66	IX—XVI	1893	December, 1893
67—90	XVII—XXVI	1905	November, 1905
91—132	XXVII—XXIX	1907	December, 1907
133—138 (including Index)	—	1908	December, 1908

The Plates are intended to be collected and bound at the end of the Volume.

A MONOGRAPH

ON THE

BRITISH FOSSIL

ECHINODERMATA

FROM

THE CRETACEOUS FORMATIONS.

VOLUME SECOND.
THE ASTEROIDEA AND OPHIUROIDEA.

BY

W. PERCY SLADEN, F.L.S., F.G.S.,

AND

W. K. SPENCER, B.A., F.G.S.

LONDON:
PRINTED FOR THE PALÆONTOGRAPHICAL SOCIETY.
1891—1908.

PRINTED BY ADLARD AND SON, LONDON AND DORKING.

THE

PALÆONTOGRAPHICAL SOCIETY.

INSTITUTED MDCCCXLVII.

VOLUME FOR 1890.

LONDON:

MDCCCXCI.

A MONOGRAPH

·ON THE

BRITISH FOSSIL

ECHINODERMATA

FROM

THE CRETACEOUS FORMATIONS.

VOLUME SECOND.
THE ASTEROIDEA.

BY

W. PERCY SLADEN, F.L.S., F.G.S., &c.,

SECRETARY OF THE LINNEAN SOCIETY.

PART FIRST.

PAGES 1—28 ; PLATES I—VIII.

LONDON:

PRINTED FOR THE PALÆONTOGRAPHICAL SOCIETY.

1891.

PRINTED BY
ADLARD AND SON, BARTHOLOMEW CLOSE.

A MONOGRAPH

ON THE

FOSSIL ECHINODERMATA

OF THE

CRETACEOUS FORMATIONS.

THE ASTEROIDEA.

INTRODUCTORY REMARKS.

It was the intention of the late Dr. Thomas Wright to have continued his magnificent series of Monographs on the British Fossil Echinodermata of the Oolitic and Cretaceous Formations, which have already appeared in the volumes of the Palæontographical Society, by the publication of a Monograph on the Cretaceous Asteroidea. With this object in view a number of plates had been prepared under Dr. Wright's directions, and some preparatory notes for the letterpress had been written, when the work was cut short by the lamented death of the author. Subsequently the Council of the Palæontographical Society did me the honour of inviting me to undertake the memoir. The plates and notes above mentioned were placed at my disposal, but the latter proved to be for the main part merely summaries or transcripts of descriptions already published, and were unfortunately unsuitable to form part of the letterpress. For the whole of the latter I am

1

therefore responsible. The plates which were drawn on stone have all been utilised, although the specimens illustrated were not in every case those which I should have selected, nor the order in which the figures are associated on some of the plates that which I should have followed. This, however, is a comparatively small matter, and the remark is not intended in any way as disparaging the excellence of the illustrations. Indeed, I would here bear unqualified testimony to the careful and accurate way in which the fossils have been delineated by Mr. A. H. Searle. His plates are monuments of patient study of morphological detail, and of exquisite technical execution as examples of lithographic drawing.

In his Monograph on the Oolitic Asteroidea, Dr. Wright gave as an introduction a general account of the structure of the main divisions of the Asteroidea then known, recent as well as fossil, with special reference to the calcareous framework; and he also gave a summary of the different systems of classification which had been formulated by previous writers on the subject. It would therefore, in my opinion, be out of place, and in a certain measure superfluous, to preface the present memoir with a similar introduction; but, as the knowledge of recent Starfishes has been considerably extended since the date of Dr. Wright's contribution, I propose to give in an appendix to this monograph my views on the classification of the Asteroidea, with special reference to the fossil forms.

At the commencement of his splendid Monograph on the Cretaceous Echinoidea—to which the present memoir is a sequel—Dr. Wright gave a valuable stratigraphical summary of the Cretaceous Formations in Britain. It is consequently altogether needless to burden the pages of the Society's publications with a repetition of these details. I shall, however, if necessary on the completion of my work, give a synopsis of the distribution in time of the various species dealt with, together with such remarks on their occurrence and associations as occasion may require.

FOSSIL ASTEROIDEA.

DESCRIPTION OF THE CRETACEOUS SPECIES.

Sub-class—EUASTEROIDEA, *Sladen*, 1886.

Order—PHANEROZONIA, *Sladen*, 1886.

Family—PENTAGONASTERIDÆ, *Perrier*, 1884.

Phanerozonate Asterids, with thick and massive marginal plates, which may be either naked, or bear granules or spiniform papillæ. Disk largely developed. Apical plates often increscent. Abactinal surface tessellate, with rounded, polygonal or stellate plates, which may be tabulate or paxilliform. Actinal interradial areas largely developed, covered with pavement-like plates, which may be naked or covered with membrane, or may bear granules or spinelets.

The family Pentagonasteridæ, as defined by Prof. Edmond Perrier[1] in 1884, was separated from a larger and more comprehensive group of genera which had been previously recognised by him[2] as constituting the family Goniasteridæ. The name Goniasteridæ was not retained for any of the groups or families into which that incongruous series of genera was divided. Previous to 1875, even the generic name of *Goniaster* had been very loosely and incorrectly applied. The vaguest notions as to the limits or characters of the genus seem to have been held. The mere form of the body, and the applicability of the significant name, irrespective of structural details, appear to have alone determined the reference of a large number of the species which have at different times borne the generic name of *Goniaster*.

---

[1] 'Nouv. Archives Mus. Hist. Nat.,' 2e série, 1884, t. vi, p. 165.
[2] 'Révis. Stell. Mus.,' p. 25 ('Archives de Zool. expér.,' 1875, t. iv, p. 289).

M. Perrier showed that none of the recent species ranked as *Goniaster* previous to 1875 had any right to be so called. He consequently employed the name in a new and restricted sense, taking the *Asterias obtusangula* of Lamarck as the type of the genus. No other species is at present known which can be regarded as congeneric with that form.

A large number of fossil Starfishes have been named as species of *Goniaster*, but none of them present characters which justify their reference to that genus in its new sense, and none of them invalidate the course taken by Prof. Perrier. It will therefore be unnecessary in the following pages to discuss in each case separately the reasons for removing the large number of species which have from time to time been ranked under the name of *Goniaster*.

Subfamily—PENTAGONASTERINÆ, *Sladen*.

PENTAGONASTERINÆ, *Sladen*. Zool. Chall. Exped., part li, Report on the Asteroidea, 1889, pp. xxxi, 262.

Pentagonasteridæ with the abactinal area paved with rounded, polygonal, or paxilliform plates. Granules or spinelets when present co-ordinated.

Genus—CALLIDERMA, *Gray*, 1847.

CALLIDERMA, *Gray*. Proc. Zool. Soc. Lond., part xv, 1847, p. 76; Ann. and Mag. Nat. Hist., 1847, vol. xx, p. 198; Synop. Spec. Starf. Brit. Mus., 1866, p. 7.

Marginal contour stellato-pentagonal. General form depressed. Disk large and flat. Rays moderately elongated and tapering. Marginal plates forming a broad border to the disk, and may be united along the median abactinal line of the ray throughout [or, in some fossil species, may be separated by one or more series of medio-radial plates, at least at the base of the ray]. The marginal plates of both series are granulated. [In recent species the supero-marginal plates bear some small papilliform spinelets on the margin where the abactinal and lateral surfaces of the plate unite; and the infero-marginal plates have a number of similar, but larger and more fully developed, spinelets irregularly distributed amongst the granulation of the actinal surface.] Abactinal area of the disk covered with small and regularly arranged plates, hexagonal in the radial areas, bearing co-ordinated granules, and some with a larger, globular, central, tubercle-like granule. Actinal interradial areas large, confined to the disk. Actinal intermediate plates large, covered with granules [and in the recent species bearing one or occasionally two compressed acute papilliform spinelets]. Armature of the adambulacral plates arranged in longitudinal series.

This genus was established by Dr. J. E. Gray for the reception of a recent Starfish, the type of which is preserved in the British Museum. It was described under the name of *Calliderma Emma*. In his remarks which follow the diagnosis, Dr. Gray observes[1] that " there is a fossil species, very like the one here described, found in the chalk, and figured in Mr. Dixon's work on the fossils of Worthing, which I propose to call *Calliderma Dixonii*." I have not been able to trace which of the fossil species is here referred to, but that is a circumstance of no great importance, as the forms figured in Mr. Dixon's work on ' The Geology of Sussex ' were described and named by the late Prof. Edward Forbes. It is interesting, however, to note that the resemblance of some of the Cretaceous forms to the genus *Calliderma* had actually been observed by the author of that genus.

Thanks to the careful study and critical insight of Mr. J. Walter Gregory, of the British Museum, a number of the examples which now form part of the National Collection have been correctly, as I think, referred to the genus *Calliderma*, and already bear that name upon the manuscript labels attached by him to the specimens.

There are, however, some differences between the fossil forms and the recent type. The most notable perhaps being the character presented by the spinulation of the marginal, the abactinal, and the actinal intermediate plates in the recent species, as compared with the fossil examples, whose state of preservation does not permit of our positively asserting whether the same character was present in their case or not. I am inclined to think that this uncertainty does not necessarily invalidate the reference of the fossil forms to the genus, and I consider it highly probable that species might exist which did not bear incipient spinelets on the plates in which they are found in the solitary existing species with which we are acquainted. The peculiar pits found upon the plates in some of the fossil examples may indicate the former presence of these spinelets, although, for my own part, I am more disposed to believe that in the majority of cases the depressions in question are structures associated with a pedicellarian apparatus. (See, for example, Pl. I, figs. 1 *a*, 1 *b*, 1 *c*, 1 *d*; Pl. III, fig. 3 *a*; Pl. V, figs. 2 *a*, 2 *b*, 2 *d*.) In other cases it is certain that little spinelets did exist, as in the tip of the ray shown in Pl. VIII, fig. 2 *a*; also, but perhaps more doubtfully, in Pl. VII, figs. 4 *a*, 4 *c*.

---

[1] ' Proc. Zool. Soc. Lond.,' part xv, 1847, p. 76; ' Synop. Spec. Starf. Brit. Mus.,' London, 1866, p. 7.

Another point of difference is to be found in some of the fossil forms which are referred in the following pages to the genus *Calliderma;* and this consists in the separation of the supero-marginal plates, at least at the base of the ray, by one or more series of medio-radial plates. It is a character whose importance is not to be under-estimated, but too little as yet is known of the morphological plasticity of the genus to justify in my opinion the separation of the forms on this ground alone. I prefer, therefore, to regard this extension of the abactinal plating as a transitional character, and I believe that this opinion is warranted by the range of plasticity observed in other genera of recent Asteroidea.

1. CALLIDERMA SMITHIÆ, *Forbes*, sp. Pl. I, figs. 1 *a*—1 *f;* Pl. VIII, figs, 2 *a*—2 *c.*

GONIASTER (ASTROGONIUM) SMITHII, *Forbes*, 1848. Memoirs of the Geological Survey of Great Britain, vol. ii, p. 474.

— — — — 1850. In Dixon's Geology and Fossils of the Tertiary and Cretaceous Formations of Sussex, London, 4to., p. 334, pl. xxii, figs. 1 and 2.

— — SMITHIÆ, *Morris*, 1854. Catalogue of British Fossils, 2nd ed., p. 80.

ASTROGONIUM SMITHII, *Dujardin and Hupé*, 1862. Hist. Nat. Zooph. Échin. (Suites à Buffon), p. 399.

GONIASTER SMITHI, *Quenstedt*, 1876. Petrefactenkunde Deutschlands, I. Abthl., Bd. iv, p. 64.

— (ASTROGONIUM) SMITHIÆ, *Forbes*, 1878. In Dixon's Geology of Sussex (new edition, Jones), p. 367, pl. xxv, figs. 1, 2, 2 *a.*

Body of large size. General form depressed. Abactinal area probably capable of slight inflation, and more or less flexible : a slight carination being present in the radial abactinal regions. Actinal surface flat. Marginal contour stellato-pentagonal, the major radius measuring rather more than twice the minor radius. Rays broad at the base and tapering gradually to the extremity. Interbrachial arcs well rounded and forming a regular curve. Margin thick, with a well-defined channel traversing the line of junction of the supero-marginal and infero-marginal series of plates, formed by the tumid character of the marginal surface of both series of plates.

The infero-marginal plates are about twenty or twenty-one in number, counting from the median interradial line to the extremity. They form a broad conspicuous border to the actinal area, the breadth of which diminishes gradually from the median interradial line to the extremity. The largest infero-marginal plates near the median interradial line measure 9 mm. in breadth, and 4 mm. in length; the length increases a little between this point and the base of the ray, where it is again 4 mm. The breadth decreases step by step from the median interradial line, and at the base of the ray is less than 4 mm., and further out the breadth of the plates is less than the length. The height of the infero-marginal plates as seen in the margin is greater than the length of the plate, the proportions near the median interradial line being as 3 : 2 approximately. The proportion of the height decreases at the extremity of the ray. The infero-marginal plates are slightly convex on their actinal surface and distinctly tumid on their marginal surface. The whole superficies is covered with small, hexagonal, closely-placed punctations, upon which granules were previously borne, probably uniform in size and compactly placed. On a number of the plates are one or more subcircular or irregular shallow concavities, quite irregular in size, position, and occurrence, which I believe to have been caused by the presence of a pedicellarian apparatus, perhaps the cavities of ordinary foraminate pedicellariæ enlarged by weathering. These are seen in Pl. I, figs. 1 a, 1 b, 1 c, 1 d. I scarcely think that they are the marks left by tubercles or enlarged granules. In the example, however, figured on Pl. VIII, fig. 2 a, small spinelets were undoubtedly present.

The adambulacral plates are broader than long, their dimensions at a short distance from the mouth being 3·25 mm. broad and 1·75 mm. long. They bear upon their surface four or five ridges, parallel or sub-parallel to the ambulacral furrow, each with five or six articulatory elevations upon which spines had previously been borne. A number of these spines are still preserved, irregularly strewed over the surface of the plates. They are short, tolerably robust, slightly flattened, slightly tapering and abruptly rounded at the tip. The longest measures about 1·5 mm. in length, or a little more; their surface is finely striate, in fact so fine that the character is perhaps mainly due to the effect of weathering upon the structural texture of the spine.

The actinal interradial areas are large and are covered with a great number of small, regular, quadrangular or rhomboid intermediate plates, which are arranged in series parallel to the ambulacral furrow, and form a compact tessellated pavement. The average size of the plates is about 2 mm. in diameter, but the plates of the series adjacent to the adambulacral plates are somewhat broader, and the plates near to the infero-marginal plates become smaller and irregular. The plates extend at the base of the ray to about the eighth infero-marginal plate, counting from the median interradial line. The surface of the plates is covered with large, rather widely spaced, hexagonal punctations—the marking left by the granules previously borne upon the plates, which appear to have been rather large and uniform.

Very few, if any, pedicallariæ appear to have been borne on the actinal intermediate plates.

The character of the mouth-plates and their armature cannot be made out in any of the examples I have examined.

The supero-marginal plates are only exposed in the marginal view of the example under description. Their height is seen to be less than that of the infero-marginal plates, and they are also rather smaller both in length and breadth; about twenty-two appear to be present between the median interradial line and the extremity of the ray. In their general character and ornamentation they resemble closely those of the infero-marginal series.

From another example, also contained in the British-Museum Collection, and bearing the registration mark " E. 2037," in which the abactinal surface is shown, the following details are supplemented.

The abactinal area is covered with small, regular, hexagonal plates or paxillar tabula, which are slightly rounded superficially, and a little bevelled on the margin of the tabulum. The surface of the tabulum is covered with punctations or marks left by the granules originally borne on the plate, and here and there small pedicellarian foramina may be seen, usually near the margin of the plate. The plates (or paxillæ) of the median radial series are broader than any of the others, the largest measuring about 2 mm. in breadth, and a little more than 1 mm. in length. The other tabula are true hexagons, measuring about 1 mm. in diameter or a trifle more, and they are arranged in longitudinal series parallel to the median radial series. Eight or nine series are present on each side of the median radial series on the disk. Opposite the eighth supero-marginal plate, counting from the median interradial line, only the median radial series and one lateral series of tabula on each side of it are present. The median radial series then extends alone and is present at the fourteenth plate, where the ray is broken in the example under description—and looks like continuing further—probably reaching to the extremity of the ray.

Dimensions.—In the type specimen (figured on Pl. I) the major radius is 98 mm., the minor radius 48 mm., the thickness of the margin from 9 to 10 mm. The breadth of a ray between the fifth and sixth infero-marginal plates measures about 28 mm.

Locality and Stratigraphical Position.—The specimen figured on Pl. I is from the Lower or Grey Chalk at Burham, in Kent. The species has also been obtained from the Upper Chalk at Brighton (Coll. Brit. Mus.); and it is sometimes found in ferric sulphide at Amberley Pit, Sussex. The fragment figured on Pl. VIII, figs. 2 a, 2 b, 2 c, is from the Lower Chalk of Dover.

History.—The type specimen, from which this species was originally described by the late Edward Forbes in 1848, formed part of the collection of Mrs. Smith, of Tunbridge Wells. It is now preserved in the British Museum. It was first figured in Dixon's ' Geology and Fossils of the Tertiary and Cretaceous Formations of Sussex,' London, 1850. Pl. I of the present memoir is a faithful drawing of the same beautiful specimen.

2. CALLIDERMA MOSAICUM, *Forbes*, sp. Pl. V, figs. 2 a—2 e; Pl. VI, figs. 1 and 2 a, b, c; Pl. VII, figs. 4 a, b, c.

GONIASTER (ASTROGONIUM) MOSAICUS, *Forbes*, 1848. Memoirs of the Geological Survey of Great Britain, vol. ii, p. 475.

—   —   MOSAÏCUS, *Forbes*, 1850. In Dixon's Geology and Fossils of the Tertiary and Cretaceous Formations of Sussex, London, 4to., p, 334, pl. xxiv, fig. 26.

—   —   MOSAICUS, *Morris*, 1854. Catalogue of British Fossils, 2nd ed., p. 80.

ASTROGONIUM MOSAICUM,   *Dujardin and Hupé*, 1862. Hist. Nat. Zooph. Échin. (Suites à Buffon), p. 399.

GONIASTER (ASTROGONIUM) MOSAÏCUS, *Forbes*, 1878. In Dixon's Geology of Sussex, (new edition, Jones), p. 367, pl. xxvii, fig. 26.

GONIASTER MOSAICUS,   *Etheridge*, 1885. In Phillips's Manual of Geology (new edition), part ii, by R. Etheridge, p. 560.

Body of large size. Disk large. Rays narrow at the base and well produced. General form depressed and thin. Abactinal area probably capable of slight inflation, and more or less flexible; some carination present in the radial abactinal regions. Actinal surface flat. Marginal contour stellato-pentagonal, the major radius measuring more than twice and a half the minor radius. Rays narrow, the supero-marginal plates being united in the median radial line. Interbrachial arcs wide and with their curvature more or less flattened, which gives a distinctly pentagonal character to the disk. Margin rather thin, and with the lateral wall perpendicular.

2

The supero-marginal plates are about twenty-eight in number, counting from the median interradial line to the extremity. (This number is taken from the fragment figured on Pl. VII, fig. 4 *a*; in the larger example drawn on Pl. V, fig. 2 *a*, twenty-two may be counted up to the place where the ray is broken.) They form a well-defined, conspicuous border, but the breadth of this is distinctly less in proportion to the size of the disk when compared with the breadth of the marginal plates in *Calliderma Smithiæ*. The largest supero-marginal plates in the specimen figured on Pl. V, fig. 2 *a*, near the median interradial line, measure 5·25 mm. in breadth and 3·25 mm. in length. The breadth diminishes very slightly as the plates approach the base of the ray, but from that part outward the length of the plates becomes much reduced—the breadth remaining the greater dimension throughout the ray.

The supero-marginal plates are comparatively flat on the abactinal surface and only slightly depressed along their margins of juncture. The general surface of the whole series has the character of sloping at a small angle to the margin of the disk, to which it gives a slightly bevelled appearance. The marginal surface of the plate is almost vertical, the junction of the abactinal and marginal surfaces is well rounded but not tumid, and there is very slight, if any, convexity on the marginal surface, at least along the disk. The height of the plates as seen in the margin is only a little greater than the length, and the diminution in height is only very trifling as the plates proceed along the ray. The whole superficies of the plates is covered with small hexagonal punctations upon which granules were previously borne. Small foraminate pedicellariæ are occasionally present here and there upon the plates; the foramen is small and oval, and is surrounded by a definite margin or lip. Sometimes more than one are present on one plate. The example figured on Pl. VII, fig. 4 *a*, is remarkable for the presence of the prominent teat-like eminences, in the centre of which the pedicellarian foramen is situated. These eminences at first sight look like tubercles for the articulation of spines (see Pl. VII, figs. 4 *a*, 4 *c*). A similar character is also seen in the example drawn on Pl. V, fig. 2 *a*, but is less strongly marked (see fig. 2 *d*).

The abactinal area of the disk is covered with small, regular, hexagonal and tetragonal plates or paxillar tabula; those in the radial areas being regularly hexagonal and larger than those in the intermediate regions, which are rhomboid, and all diminish in size as they approach the margin. The abactinal plates or paxillæ do not appear to extend beyond the twelfth supero-marginal plate, counting from the median interradial line; the supero-marginal plates of the two sides of the ray meeting in the median radial line beyond this point. The plates or paxillæ of the median radial series are larger and broader than any of the others; they are succeeded on each side by five or six longitudinal series of hexagonal plates, those of the second or third series from the median series measuring about 1·5 mm.

in diameter. The remaining plates which occupy the intermediate areas are tetragonal or rhomboid. All the plates have their surface marked with rather widely-spaced punctations—the impressions of the granules previously present. Small foraminate pedicellariæ are also frequently present here and there, usually near the margin of the plate.

The madreporiform body is flat, distinct, and polygonal in outline; it is situated near the centre of the disk. Its surface is marked by fine straight striæ, which radiate regularly centrifugally from the centre to the margin (see Pl. V, fig. 2 e).

Other specimens show that the infero-marginal plates in this species are more nearly subequal to the supero-marginal series than in *Calliderma Smithiæ*, that the actinal intermediate plates are relatively larger than in that species and a good deal larger than the abactinal paxillar plates or tabula. The actinal intermediate plates originally bore granules only, judging from the character of the punctations with which their surface is ornamented. A fragment belonging to the British Museum Collection (which bears the register number "E 373"), in which the spines that formed the armature of the adambulacral plates are preserved, indicates that these spines are smaller, shorter, and perhaps more numerous than in *Calliderma Smithiæ*.

In the example drawn on Pl. VI, fig. 2 a, the supero-marginal plates are preserved, but the whole of the abactinal plating has been removed, leaving exposed the inner surface of some of the actinal intermediate plates and the adambulacral plates. Magnified details of these plates are given, and they represent the characters of the structures preserved better than any verbal description.

Dimensions.—The large example figured on Pl. V, fig. 2 a, has the following measurements :—Major radius 82 + mm. (all the rays are broken and imperfect, and the full dimensions cannot therefore be given); minor radius 36 mm.; thickness of the margin about 8 mm. Breadth of a ray between the eighth and ninth supero-marginal plates about 15 mm., or a trifle more.

Locality and Stratigraphical Position.—The example figured on Pl. V, fig. 2 a, is labelled from the Lower Chalk, but the locality is not recorded. It formed part of one of the old collections preserved in the British Museum. Other examples in the British Museum are from the Grey Chalk or Chalk Marl of Dover, from the Lower Chalk of Glynde in Sussex, and from the Lower Chalk of Amberley Pit, Arundel. There is also a magnificent specimen in the Museum of Practical Geology, Jermyn Street, from the Lower Chalk of Dover.

3. CALLIDERMA LATUM, *Forbes*, sp.   Pl. II, figs. 1 *a*—1 *e*, 2 *a*—2 *d*; Pl. III, figs. 1 *a*—1 *e*, 2 *a*, 2 *b*, 3 *a*, 3 *b*.

GONIASTER (ASTROGONIUM) LATUS, *Forbes*, 1848. Memoirs of the Geological Survey of Great Britain, vol. ii, p. 474.

—   —   —   —   1850. In Dixon's Geology and Fossils of the Tertiary and Cretaceous Formations of Sussex, London, 4to., p. 333, pl. xxiii, figs. 4, 5.

—   —   —   *Morris*, 1854. Catalogue of British Fossils, 2nd ed., p. 80.

ASTROGONIUM LATUM,   *Dujardin and Hupé*, 1862. Hist. Nat. Zooph. Échin. (Suites à Buffon), p. 399.

GONIASTER (ASTROGONIUM) LATUS, *Forbes*, 1878. In Dixon's Geology of Sussex (new edition, Jones), p. 367, pl. xxvi, figs. 4, 5.

Body of large or moderate size. General form depressed. Abactinal surface probably capable of some degree of inflation. Actinal surface flat. Marginal contour stellato-pentagonal, the major radius probably not exceeding the minor radius by more than one half. Rays narrow at the base, short, not greatly produced, and probably tapered to a pointed extremity. Interbrachial arcs very wide and flattened, which gives a strongly marked pentagonal outline to the disk. Margin of uniform thickness.

The infero-marginal plates are more than sixteen in number, counting from the median interradial line to the extremity (the tip of the ray being broken in all the specimens examined). They form a remarkably broad margin to the actinal area of the disk, which diminishes rather rapidly in width at the base of the rays, and then slightly to the extremity. The largest infero-marginal plates near the median interradial line measure about 13 mm. in breadth and 4·5 mm. in length. The length is nearly uniform throughout, or at any rate till well out on the free part of the ray; but the breadth diminishes until the plates at the base of the ray are 6·5 mm., and the diminution proceeds to a certain extent along the ray. The infero-marginal plates are slightly convex along their line of breadth, by which means the separate plates are distinctly marked. They are well rounded at the junction of the actinal and lateral surfaces, and are slightly tumid in the margin. The outline of their inner or adcentral edge is also rounded. The height of the infero-marginal plates as seen in the margin is a little greater than their length. The height of the supero-marginal plates is, however, somewhat greater.

The whole superficies of the plates is covered with circular punctations of irregular size rather than hexagons, as in the other forms, and the irregularity caused by the presence of larger punctations here and there is remarkable. This character seems to indicate the former presence of an irregular-sized granulation.

The supero-marginal plates are similar in character to the infero-marginal series, but the large irregular punctations are larger and more numerous.

The adambulacral plates are broader than long, and they bear upon their surface five or six ridges parallel or subparallel to the ambulacral furrow, each with prominent well-defined granulations or elevations, all uniform and closely placed, upon which the adambulacral armature of spines was previously borne (see Pl. II, fig. 1 d; Pl. III, fig. 2 b).

The actinal interradial areas are large, and are covered with comparatively large polygonal and rhomboid intermediate plates, which are arranged in series parallel to the ambulacral furrow, and originally formed a compact tessellated pavement. In a number of the fossils of this species, however, these plates are often separated and displaced, which leads to the inference that in life the plates were not so intimately connected as in other species, and that membrane or con-nective tissue was more largely developed. The one or more series of plates adjacent to the adambulacral plates are much larger than the others, and none of the intermediate plates extend beyond the base of the ray. The surface of the plates is covered with large, irregular, and rather deeply sunken pits, the character of which leads to the inference that the granulation originally present was also somewhat irregular in size and coarse in character (see Pl. II, fig. 1 c; Pl. III, fig. 1 e).

In some examples (notably in that figured on Pl. III, fig. 2 a) small oval fora-minate pedicellariæ, distinctly lipped at the margin of the foramen, are present on the actinal intermediate plates.

The mouth-plates are elongate, about three times as long as broad, triangular in shape, with the two outer sides subequal. Their surface is covered with large, coarse, irregular, tuberculose elevations (see Pl. II, fig. 1 e), which suggest the inference that the armature of the mouth-plates consisted of large, irregular, papilliform granules.

In some examples a portion of the actinal floor has been removed, exposing the inner surface of the abactinal floor. In these cases the stellate base of the abac-tinal plates or paxillæ are seen (see Pl. II, figs. 2 a, 2 d; Pl. III, figs. 3 a, 3 b). It will be noticed that there is a difference in the form of the stellate bases in these examples, which may indicate a specific or varietal difference, but I do not feel justified from this character alone in recognising either of these fragments as the type of a distinct species. More material is needed before such a step would be warranted.

Dimensions.—The large example figured on Pl. II, fig. 1 *a*, has a major radial measurement of from 80 to 95 mm. or more, with a minor radius of about 52 mm. The breadth of the ray between the sixth and seventh infero-marginal plates, counting from the median interradial line, is about 17 or 18 mm.

Locality and Stratigraphical Position.—This species appears to be confined to the Lower Chalk. Examples have been collected from Washington, Amberley, Southerham, and Glynde, in Sussex. Also from the Lower Chalk of Folkestone, and the Chalk Marl of Dover.

History.—Two examples of this species were first figured by Forbes in Dixon's 'Geology and Fossils of the Tertiary and Cretaceous Formations of Sussex,' London, 1850, pl. xxiii, figs. 4 and 5. Both these specimens are now preserved in the British Museum. One example, which is from Amberley, is drawn on Pl. III, fig. 1 *a*. The other, which is from Washington, is accurately represented by fig. 3 *a* of the same plate.

Variations.—In addition to the difference noted above in the form of the stellate bases of the abactinal plates or paxillæ, other minor differences may be observed. In some examples the breadth of the border formed by the marginal plates on the disk area is not relatively so great as in other examples, and the proportions of length to breadth, as well as the amount of tumidity of the component plates, are subject to variation. In some examples, again, the irregularity in the granulation of the marginal plates, arising from the former presence of coarser granules interspersed amongst the average granulation, is more marked than in others. These differences will be more readily noticed by turning to the figures given on Pl. II and Pl. III than by a lengthy verbal description. Some of the examples come from different beds and different localities—circumstances which I consider to be sufficient to account for the variation.

Genus—NYMPHASTER, *Sladen,* 1885.

NYMPHASTER, *Sladen.* In Narr. Chall. Exp., 1885, vol. i, p. 612 ; Zool. Chall. Exped., part li, Report on the Asteroidea, 1889, p. 294.

Disk large and flat. Rays elongate, slender, tapering, and almost square in section. Marginal plates forming a broad border to the disk, and either united along the median abactinal line of the ray throughout, or separated only by a single series of medio-radial plates. The marginal plates of both series are granulated, and bear no spines (normally, but occasional incipient spinelets may be present). Abactinal area of the disk covered with large and regularly arranged plates, those in the radial areas well separated, usually hexagonal, more or less tabulate and paxilliform, and frequently bearing an excavate or entrenched pedicellaria. Actinal interradial areas large, confined to the disk. Actinal intermediate plates well defined, covered with uniform granules, and occasionally bearing pedicellariæ. Armature of the adambulacral plates arranged in longitudinal series. Madreporiform body exposed and situated within one third of the distance from the centre to the margin. Large entrenched pedicellariæ are frequently present on the marginal plates in some species.

This genus includes a number of recent species brought to light by the deep-sea explorations of late years. It has been found in the Atlantic, the Pacific, and the Eastern Archipelago. The Atlantic species pass into the abyssal zone, but those inhabiting the Pacific and Eastern Archipelago do not, so far as at present known, extend beyond the continental zone, or in other words they live in depths of less than 500 fathoms.

The structure and character of the Cretaceous species described in the following pages, so far as they can be made out from the fragmentary condition of the fossils, appear to me to warrant their inclusion in the genus *Nymphaster*.

1. NYMPHASTER COOMBII, *Forbes*, sp. Pl. VII, figs. 1—3 ; Pl. VIII, figs. 1 *a*, 1 *b*.

GONIASTER (ASTROGONIUM) COOMBII,	*Forbes*, 1848.	Memoirs of the Geological Survey of Great Britain, vol. ii, p. 474.
— — —	— 1850.	In Dixon's Geology and Fossils of the Tertiary and Cretaceous Formations of Sussex, London, 4to., p. 334, pl. xxiii, fig. 6.
— — —	*Morris*, 1854.	Catalogue of British Fossils, 2nd ed., p. 80.
ASTROGONIUM COMBII,	*Dujardin and Hupé*, 1862.	Hist. Nat. Zooph. Échin. (Suites à Buffon), p. 399.
GONIASTER (ASTROGONIUM) COOMBII,	*Forbes*, 1878.	In Dixon's Geology of Sussex (new edition, Jones), p. 367, pl. xxvi, fig. 6.

Body of medium size. Disk moderately large. Rays well produced, rather broad at the base and tapering to the extremity. General form depressed and thin. Marginal contour stellato-pentagonal, the major radius measuring more than twice and a half the minor radius. Marginal plates broad, the supero-marginal series of the two sides of the ray meeting in the median radial line. Interbrachial arcs deeply indented and well rounded. Margin rather thin.

The infero-marginal plates are more than fifteen in number, counting from the interradial line to the extremity. They form a broad conspicuous border to the actinal area, which is relatively broad in proportion to the size of the disk. The largest infero-marginal plates near the median interradial line measure about 5·5 mm. in breadth, and about 2·5 to 2·75 mm. in length. The breadth decreases slightly from this point as the plates approach the base of the ray, and then much more rapidly, the plates on the outer part of the ray having the length considerably in excess of the breadth. The plates are tumid and roundly bevelled at the lateral edges, but are flatly rounded at the margin of the disk, and without tumidity there. The whole superficies of the plate is covered with large, rather deeply depressed, hexagonal punctations, closely placed, which give somewhat of a honeycomb appearance to the plate (see Pl. VIII, fig. 1 b). These are the marks left by the granules previously borne upon the plate. Upon a number of the plates in the example figured on Pl. VIII, fig. 1 a, the granules are still preserved in situ. They are large and closely placed. The punctations, and consequently the granules, in this species are coarser than in any of the other Cretaceous forms known to me. I have not been able to assure myself of the presence in this example of any pedicellariæ on the infero-marginal plates.

The adambulacral plates are broader than long, except on the outer part of the ray, and their armature appears to have consisted of five or six regular series of spinelets. This is indicated by the presence upon the surface of the plate of that number of ridges, running parallel or subparallel to the ambulacral furrow, each having four or five articulatory elevations and intervening pits upon which spinelets had previously been borne. The spinelets were probably short, and similar to those described in *Calliderma Smithiæ* and *Calliderma mosaicum*, but I have not found any preserved in specimens which I consider to be undoubted examples of *Nymphaster Coombii*.

Dimensions.—In the type specimen, figured on Pl. VIII, fig. 1 a, the major radius is more than 56 mm., and the minor radius 23 mm. The breadth of the ray between the fourth and fifth infero-marginal plates measures about 15 mm.

Locality and Stratigraphical Position.—The specimen figured is from the Lower Chalk of Balcombe Pit, Amberley. The species has also been obtained from the

Lower Chalk of Glynde, Sussex; as well as from the Lower Chalk of Dover and the Isle of Wight. Other specimens of *Nymphaster*, as to the reference of which to *N. Coombii* I am more or less doubtful, which show certain differences in structural details, are from the Grey Chalk of Folkestone and Dover, and from the Lower Chalk of Betchworth. Several examples in the Museum of Practical Geology, Jermyn Street, are labelled from the " Upper Chalk," but I am inclined to think that their reference to that horizon is more or less doubtful.

History.—The type of this species was found by Mr. G. Coombe at Balcombe Pit, Amberley, and formed part of Mr. Dixon's collection. It is now preserved in the British Museum. It was first figured by Edward Forbes in Dixon's ' Geology and Fossils of the Tertiary and Cretaceous Formations of Sussex,' London, 1850, pl. xxiii, fig. 6. The same specimen is carefully represented on Pl. VIII, figs. 1 *a*, 1 *b* of this memoir.

Doubtful Examples of this Species.—Three specimens are figured on Pl. VII, which I only place provisionally and with very great doubt under this species. I do not, therefore, at present propose to describe them in detail, or to definitely assign the characters they present as supplementary to those already given as belonging to *Nymphaster Coombii.*

1. An example from the Lower Chalk of Betchworth, in which a portion of the actinal surface is preserved (Pl. VII, figs. 1 *a*—1 *e*). This specimen shows large infero-marginal plates somewhat longer in proportion to their breadth than in the type specimen, and their surface is covered with an extremely fine uniform punctation. The latter character is altogether unlike that of examples which I consider to be true forms of *Nymphaster Coombii.* But from this character alone, which recent forms show to be one subject to considerable variation, I shrink from taking any more definite step, at least until further material is available for study. This example has some of the adambulacral plates and actinal intermediate plates well preserved. The adambulacral plates (see Pl. VII, fig. 1 *c*) conform to the description given above. The actinal intermediate plates are rhomboid in form, and their surface is covered with deep, large, well-spaced pits, which indicate the former presence of a coarse uniform granulation. These plates are shown on Pl. VII, fig. 1 *e*. The margin of this example is quite characteristic of *Nymphaster Coombii.* The infero-marginal plates are seen to be low and more or less bevelled or sloping towards the margin; whilst the supero-marginal plates are relatively rather higher and more abruptly bent at the junction of the actinal and lateral surfaces (see Pl. VII, fig. 1 *b*).

2. This is a badly preserved specimen from the Grey Chalk of Folkestone, in which nothing but the supero-marginal plates and the general outline are available

3

for determination (see Pl. VII, figs. 2 *a*, 2 *b*). The marginal plates resemble in character those of the specimen just mentioned, and they are like them covered with a very fine punctation, unlike that of the typical *Nymphaster Coombii*. There are also fewer plates in that portion of the interbrachial arc which may be said to belong to the disk than in *Nymphaster Coombii*, but as the example is smaller, this may probably be only a question of age ; or it may, like the punctation of this and the preceding example, be attributed to variation, which I am disposed to consider a not improbable reason for the differences, when regard is had to the horizon from which the fossils were obtained, and consequently the changed conditions of existence in which those Asterids probably lived.

3. This specimen (figured on Pl. VII, figs. 3 *a*, 3 *b*) is from the Lower Chalk of Glynde, Sussex, and I consider that its reference to *Nymphaster Coombii* is less doubtful than that of either of the two preceding examples. The fragment represents a portion of the abactinal surface. The supero-marginal plates are large, and are covered with the characteristic coarse punctation of *Nymphaster Coombii* (see Pl. VII, fig. 3 *b*). The supero-marginal plates of the two sides of a ray meet in the median radial line from the very base of the ray, distinctly characteristic of the genus *Nymphaster*. Comparing this example with the typical form of the species, there appear to be a much smaller number of supero-marginal plates in the interbrachial arc belonging to the true disk, and on these grounds I hesitate from accepting it positively as an undoubted example of this species until further material is forthcoming to throw light upon the amount of plasticity which may be accredited to this species.

2. NYMPHASTER MARGINATUS, *Sladen*. Pl. VIII, figs. 4 *a*, 4 *b*.

Body of medium size. General form depressed. Marginal contour stellato-pentagonal. Rays well produced, rather broad at the base, and tapering gradually to a pointed extremity. Interbrachial arcs deep and rounded, the sweep of the curve from the tip of one ray to the tip of the neighbouring ray being of a paraboloid character. Margin rather thin.

The supero-marginal plates form a broad and massive border to the abactinal area of the disk. There are six plates on each side of the disk counting from the base of one ray to the base of the neighbouring ray. All the succeeding plates along the ray meet the corresponding plate of the opposite side of the ray in the median radial line. The abactinal surface of the ray is thus occupied entirely by the supero-marginal plates throughout its length.

All the supero-marginal plates are of uniform height, excepting the normal diminution towards the extremity of the ray ; and all have the breadth greater than the length. The plates on each side of the median interradial line measure about 4 mm. in breadth and about 2 mm. in length ; and this length is maintained with very slight diminution until about midway between the base and the extremity of the ray, where the length is 1 75 mm., and the breadth is between 2·75 and 3 mm. Sixteen supero-marginal plates are preserved in the longest ray of the specimen under description, counting from the median interradial line to the broken extremity. A few plates are apparently missing. Measured in the margin, the height of the plates is about 2·5 mm.

All the supero-marginal plates are distinctly convex in the direction of their median line of breadth, by which each plate is very clearly marked out, and a highly ornate character is given to the species. The plates are also tumid and well rounded on their marginal surface. The whole surface of the plates is covered with rather large, widely spaced punctations or pits, which have a peculiarly isolated appearance, unlike that of any other species (see Pl. VIII, fig. 4 b). I have not detected the presence of any pedicellariæ upon this example.

The remains of a few isolated plates are preserved on the abactinal area of the fossil figured. They are all small and out of position, and are not available for description.

Dimensions.—The specimen figured on Pl. VIII, fig. 4 a, has a minor radius of about 12 mm.; and the longest fragment of a ray preserved measures about 35 mm. The extremity is wanting. The thickness of the margin is 4·5 mm. The breadth of the ray at the base between the third and fourth supero-marginal plates counting from the median interradial line is from 8 to 8·5 mm.

Locality and Stratigraphical Position.—The example described is from the Upper Chalk near Bromley. It is preserved in the British Museum, and bears the registration number 35,484.

3. NYMPHASTER OLIGOPLAX, *Sladen.* Pl. VIII, figs. 3a, 3b.

Body of medium size. General form depressed and thin. Marginal contour stellato-pentagonal. Rays narrow at the base and produced. Interbrachial arcs wide and rounded. Margin thin.

The supero-marginal plates form a broad border to the abactinal area of the disk. There are only three (or possibly four) supero-marginal plates between the median interradial line and the base of the ray—that is to say, about six plates on each side of the disk. The fourth (or perhaps the fifth) plate counting from the median interradial line, and all the succeeding plates along the ray, appear normally to meet the corresponding plate of the opposite side of the ray in the median radial line. The abactinal surface of the ray is thus occupied entirely by the supero-marginal plates throughout its length. In one of the rays preserved there appear, however, to be traces of a few abactinal plates which interfere with the union of the supero-marginal plates in the median radial line near the base of the ray. As to how far this is normal I am unable to say.

All the supero-marginal plates are comparatively low and flat. The plates on each side of the median interradial line are 3·5 mm. in breadth, and from 3 to 3·5 mm. in length, and are thus practically square. Their abactinal surface is slightly convex; and their height as seen in the margin is less than the length, and the abactinal surface is gradually bevelled to the inferior margin which abuts upon the infero-marginal plates. The other plates which form the border of the disk-area are of the same size and character as those adjacent to the median inter-radial line. The supero-marginal plates along the ray have the breadth greater than the length, the fifth plate counting from the median interradial line measuring about 3·75 mm. in breadth and 2·5 mm. in length. Their character is similar to that of the plates above described. The surface of the plates is covered with small well-spaced punctations, and there is a distinct smooth border on the inner and two lateral margins of each plate on which no punctations or pits are present.

Large trench-like pedicellariæ, which are nearly the length of the plate, are present in this species; they occur more frequently on the infero-marginal plates than on those of the superior series; in fact, only one or two are present on the latter series of plates in the example under description.

No other portions of this fragment are available for description.

Dimensions.—The fragment figured in Pl. VIII, fig. 3 *a*, has a minor radius of about 15 mm. The longest portion of the major radius preserved is about 33 mm., and the ray is broken off abruptly. The thickness of the margin is between 4 and 5 mm. The breadth of the ray at the base is about 8 mm.

Locality and Stratigraphical Position.—The fragment described, which is, unfortunately, all that I have seen, is from the Upper Chalk of Bromley. It is preserved in the British Museum, and bears the registration number 40,178.

Remarks.—The character of the marginal plates, as regards both their form and their ornamentation, as well as the presence of the peculiar pedicellariæ, and indeed the whole facies of this fossil, lend strong support to the presumption that this species may ultimately need to be placed in a distinct genus, but I do not feel warranted in taking that step on the basis of such scanty material.

Genus—PYCNASTER, *Sladen.*

Disk relatively small and pentagonal. Abactinal surface more or less convex, and was probably somewhat inflated during life. Margin thick, and highest in the region of the disk. Rays elongate, narrow, and robust. Marginal plates forming a broad border to the disk, and united along the median abactinal line of the ray throughout. The marginal plates are high and very robust, those of the superior series being prominently convex abactinally in the median line of breadth and height, which imparts a well-rounded character to the ray. The marginal plates of both series are finely granulated, and probably bore no spines. Actinal intermediate plates large, covered with uniform granules. Armature of the adambulacral plates arranged in longitudinal series. Foraminate pedicellariæ with radiating channels may be present on the marginal plates.

The fragmentary state of the fossils which I have referred to this type unfortunately does not permit of a complete diagnosis of the genus being drawn up. The characters above given appear, however, to me to be sufficient to warrant the recognition of the possessors of them as the representatives of a distinct genus. The small high disk, the massive convex marginal plates, and the large actinal intermediate plates, together with the form of the rays, produce a facies alone sufficient to stamp its individuality, irrespective of other details of structure.

1. PYCNASTER ANGUSTATUS, *Forbes,* sp. Pl. IX, figs. 1 *a,* 1 *b.*

GONIASTER (ASTROGONIUM) ANGUSTATUS, *Forbes,* 1848. Memoirs of the Geological Survey of Great Britain, vol. ii, p. 474.

—   —   —   — 1850. In Dixon's Geology and Fossils of the Tertiary and Cretaceous Formations of Sussex, London, 4to., p. 335, pl. xxiii, fig. 10.

GONIASTER (ASTROGONIUM) ANGUSTATUS, *Morris*, 1854. Catalogue of British Fossils, 2nd ed., p. 80.

ASTROGONIUM ANGUSTATUM, *Dujardin and Hupé*, 1862. Hist. Nat. Zooph.Échin.(Suitesà Buffon), p. 399.

GONIASTER (ASTROGONIUM) ANGUSTATUS, *Forbes*, 1878. In Dixon's Geology of Sussex (new edition, Jones), p. 368, pl. xxvi, fig. 10.

Disk of medium size or relatively small and pentagonal. Rays elongate, narrow, robust, and, though tapering, nearly uniform in breadth throughout. Marginal contour stellate. Interbrachial arcs more or less flattened, which emphasises the pentagonal outline of the disk. Margin thick and robust, much highest in the region of the disk. Abactinal surface more or less convex, and was probably somewhat inflated during life. Actinal intermediate plates very large, covered with uniform granules.

The supero-marginal plates are thick and massive, and they form a high and broad border to the disk. There are only three supero-marginal plates between the median interradial line and the base of the ray—that is to say, six plates on each side of the disk. The fourth plate counting from the median interradial line, and all the succeeding plates along the ray, meet the corresponding plate of the opposite side of the ray in the median radial line. The abactinal surface of the ray is thus occupied entirely by the supero-marginal plates throughout its length.

The supero-marginal plates which form the border of the disk are much larger in the direction of height than any of the others. The plates on each side of the median interradial line are about 4·75 mm. in breadth as seen on the abactinal surface, and about the same measurement in length. They are convex abactinally, and well rounded at the junction of the abactinal and lateral surfaces. Measured in the margin their height is 8 mm., and their lateral surface (which forms the vertical wall of the margin) is distinctly convex or pulvinate, but to a less degree than their abactinal surface.

The supero-marginal plates of the ray are not so high as those of the disk, although their height is greater than their length. The height of the sixth plate from the median interradial line is about 5 mm. Their abactinal and lateral surfaces form together a true segment of a circle, and this imparts a well-rounded character to the ray. The plates are deeply bevelled at their junction with the adjacent plates, and consequently distinctly pulvinate in the median line of breadth and height. The surface of the plates is covered with minute punctations, but these are so extremely faint that they are seen with difficulty. They are probably weather-worn in the example under notice.

The infero-marginal plates, as seen in the direct lateral view of the margin, are much smaller in height than the supero-marginal series in the type specimen. The plates which form the margin of the disk are higher than long, the height being about 5 mm. and the length about 3·25 mm. in those adjacent to the median interradial line; the succeeding plates on the margin of the disk are each less in height than the preceding plate, the third or fourth plate, counting from the median interradial line, having the height and length about equal. The infero-marginal plates along the ray have the length greater than the height. The surface of the infero-marginal plates resembles that of the superior series in the character of its punctation.

Traces of small excavate pedicellariæ are present on occasional plates, but these appear to have been very few in the example under description.

On the abactinal surface of the disk a few isolated and displaced plates are present. Some of these seem rather thick and tuberculous in character, but the state of the preservation of this part of the fossil is unfortunately quite unfitted for description.

There is a fine fragment of this species preserved in the Museum of Practical Geology, Jermyn Street, from the Upper Chalk of Bromley, which shows part of the actinal surface. The infero-marginal plates in the disk are very high in this example, and five of them in an interbrachial arc bear a small foraminate pedicellaria. This is situated near the upper margin of the plate, about equidistant from that margin and the lateral margins of the plate, and consists of a small round foramen situated in the middle of a very shallow concavity, and with five or six faint channels radiating from the foramen to the margin of the concavity, gradually thinning and dying out there. The channels radiate like the spokes of a wheel, or a five-rayed star, and produce a facies unlike that of any other pedicellarian apparatus with which I am acquainted. The actinal intermediate plates are very large, and not more than three series are present. The plates of the series next to the adambulacral plates are much larger than the others, and are broader than long. The adambulacral plates are broader than long, and their surface is marked with three or four ridges parallel to the furrow, upon which spinelets were previously borne. The furrow series consists of about five spinelets. A few of these spinelets are preserved, and they are rather short, cylindrical, and slightly tapering. The mouth-plates are very small and narrow.

Dimensions.—The specimen figured on Pl. IX, fig. 1 *a*, has a minor radius of about 23 mm. The longest portion of a major radius preserved is 53 mm.; the ray is broken abruptly, and there is very slight diminution in the breadth at the broken extremity as compared with the breadth at the base; there would appear to be every indication that only a small part of the ray is preserved.

The thickness of the margin at the median interradial line is 13 mm., and at the base of the ray 8·5 mm. The breadth of the ray at the base is about 12 mm.

Locality and Stratigraphical Position.—The example above described, which has been drawn on Pl. IX, fig. 1 a, was obtained from the Upper Chalk in Kent, but unfortunately the exact locality is unknown. It is preserved in the British Museum. A fine fragment preserved in the Museum of Practical Geology, Jermyn Street, was obtained from the Upper Chalk of Bromley. The species has also been found in the Upper Chalk of Sussex.

History.—The type of this species was first described by Forbes under the name of *Goniaster (Astrogonium) angustatus*, and was afterwards figured by him in Dixon's 'Geology and Fossils of the Tertiary and Cretaceous Formations of Sussex,' London, 1850, pl. xxiii, fig. 10. That illustration does not, however, give a good idea of the facies of the species.

Genus—PENTAGONASTER, *Linck*, 1733.

PENTAGONASTER,	*Linck.*	De Stellis marinis, 1733, p. 20.
—	*Schülze.*	Betrachtung der versteinerten Seesterne und ihrer Theile, Warschau u. Dresden, 1760, p. 50.
GONIASTER (pars),	*L. Agassiz.*	Prod. Mon. Radiaires, Mém. Soc. Sci. Nat. Neuchatel, 1835, t. i, p. 191.
ASTROGONIUM (pars),	*Müller and Troschel.*	System der Asteriden, 1842, p. 52.
GONIODISCUS (pars),	*Müller and Troschel.*	Ibid., 1842, p. 57.
HOSIA (pars),	*Gray.*	Ann. and Mag. Nat. Hist., 1840, vol. vi, p. 279.
TOSIA,	*Gray.*	Ibid., 1840, vol. vi, p. 281.

Body depressed and pentagonal in contour, or with the rays slightly produced. Marginal plates smooth or granular, ordinarily few in number. Supero-marginal plates form a broad border to the disk, and, when the ray is produced, are separated throughout by abactinal plates. Abactinal area covered with rounded or polygonal plates, which may either be smooth or bear co-ordinated granules. Actinal intermediate plates and infero-marginal plates smooth or granulose, devoid of prominent spinelets.

Much diversity of opinion has existed, unnecessarily it seems to me, as to the character and limits of this genus. Two species were originally referred to *Pentagonaster* by its founder. The type of one of these is now lost, and its identification rests only on surmise. The second species, however, *Pentagonaster semilunatus*, is a well-known and widely distributed recent form, about which there is no doubt. I therefore consider that this form has every claim to be regarded as the type of the genus. The existing species of *Pentagonaster* are found in the Atlantic, the Pacific, the Indian and the Southern Oceans, and in the Eastern Archipelago; and the bathymetrical range of the genus extends from 20 to 1500 fathoms or more.

1. PENTAGONASTER LUNATUS, *Woodward*, sp. Pl. IV, figs. 1 *a*—1 *c*.

ASTERIAS LUNATUS, *Woodward*, 1833. An Outline of the Geology of Norfolk,
 p. 52, pl. v, fig. 1.
TOSIA LUNATA, *Morris*, 1843. Catalogue of British Fossils, p. 60.
 — — *Bronn*, 1848. Index Palæontologicus, Nomenclator, p. 1274.

Body of medium size. General form depressed. Abactinal and actinal areas flat. Marginal contour stellato-pentagonal, the major radius measuring nearly twice the minor radius. Rays short and moderately produced, rather narrow at the base and tapering to the extremity. Interbrachial arcs deeply indented and well rounded.

The infero-marginal plates are twelve (or more) in number, counting from the median interradial line to the extremity. They form a broad border to the actinal area of the disk, the breadth of which diminishes rather rapidly plate by plate as they recede from the median interradial line. The largest infero-marginal plates adjacent to the median interradial line measure about 5·25 mm. in breadth and about 3 mm. or a little more in length. The length and breadth decrease as each plate proceeds outward until about midway on the ray, where these dimensions are subequal. On the outer part of the ray the length is greater than the breadth. The infero-marginal plates are distinctly convex on their actinal surface in the direction of the median line of breadth, and are slightly tumid at the margin. Their surface is covered with small, uniform, closely placed, and deeply sunken moniliform punctations, upon which small granules were previously borne, probably uniform in size and closely placed (see Pl. IV, fig. 1 *c*). I am not aware that traces of any pedicellariæ have been detected on these plates.

The adambulacral plates are small and oblong, and bear on their surface ridges

4

of alternating granuliform eminences and depressions, upon which the spinelets constituting the armature of the adambulacral plates were originally borne.

The actinal interradial areas are small, and are covered with regular pentagonal or rhomboid intermediate plates, which are arranged in series parallel to the ambulacral furrow, and form a compact, mosaic-like pavement. The actinal intermediate plates are moderately large in relation to the size of the disk. The plates of the series adjacent to the adambulacral plates are sensibly larger than any of the others, and the plates of the next series are also larger than those which form the rest of the pavement. Near the infero-marginal plates the actinal intermediate plates become small and more or less irregular. The intermediate plates extend at the base of the ray to about the fifth infero-marginal plate, counting from the median interradial line. The surface of the plates is covered with rather large, widely spaced, and deeply sunken punctations, upon which granules were previously borne, and these would appear to have been comparatively large in size and uniform (see Pl. IV, fig. 1 *b*).

From what is visible of the margin of this example it is seen that the supero-marginal series of plates are nearly of the same height as the infero-marginal series, and are similar in structure.

Unfortunately no other portions of this fragment are available for description.

Locality and Stratigraphical Position.—The specimen upon which this species was founded was collected by Mr. Samuel Woodward, from the Upper White Chalk, near Norwich.

Dimensions.—In the type specimen (figured on Pl. IV, fig. 1 *a*) the major radius is about 35 mm., and the minor radius about 18 mm. Breadth of the ray between the fifth and sixth infero-marginal plates about 10 mm.

History.—The type specimen was figured by Woodward in his ' Outline of the Geology of Norfolk,' pl. v, fig. 1, and is now preserved in the collection of the Norfolk and Norwich Museum. It was kindly lent by the committee of that institution to Dr. Wright for the purpose of this monograph. It has been carefully drawn on Pl. IV, figs. 1 *a*—1 *c*. An admirably executed cast of this specimen is in the British Museum. I am not at present aware of the existence of any other examples of this rare form.

Remarks.—The example referred by Forbes to this species, and figured by him in Dixon's ' Geology and Fossils of the Tertiary and Cretaceous Formations of Sussex,' London, 1850, pl. xxiii, fig. 9, belongs to a distinct species, which I have named *Pentagonaster megaloplax*. A number of other specimens in other collections have, following Forbes, been erroneously referred to *Pentagonaster lunatus*, which are in reality examples of *Pentagonaster megaloplax*. This is unfortunate, for the latter form has thus become comparatively well known under the name of *Pentagonaster lunatus*, a name which they must now cease to bear, as the real *Pentagonaster lunatus* is quite a different form, and there is no doubt whatever either as to the type (which is preserved in Norwich) or the priority. The differences between the two species will be further noticed under the description of *Pentagonaster megaloplax*.

2. PENTAGONASTER MEGALOPLAX, *Sladen*. Pl. IV, figs. 2—4.

GONIASTER (ASTROGONIUM) LUNATUS, *Forbes*, 1850. In Dixon's Geology and Fossils of the Tertiary and Cretaceous Formations of Sussex, London, 4to., p. 353, pl. xxiii, fig. 9 (non *Asterias lunatus*, Woodward, 1833).

Body of medium size. General form depressed. Abactinal and actinal areas flat. Marginal contour stellato-pentagonal, the major radius measuring a little more than once and a half the minor radius. Rays short and not greatly produced, tapering gradually to the extremity. Interbrachial arcs regularly rounded, curving gradually from the tip of one ray to that of the adjacent ray, which gives a distinctly lunate character to the disk. Margin of uniform thickness.

The infero-marginal plates are only five or rarely six in number, counting from the median interradial line to the extremity. They form a very broad border to the actinal area of the disk in relation to its size, and the breadth is maintained until near the extremity. The largest infero-marginal plates adjacent to the median interradial line measure 7 mm. in breadth, and about 6·5 mm. or nearly 7 mm. in length; they are consequently almost square. The proportion of breadth diminishes in the succeeding plates as they recede from the median interradial line. The infero-marginal plates have a more or less pulvinate appearance actinally, consequent on being rounded or bevelled at the edges; and they are slightly tumid in the margin. Their whole actinal surface is covered with large, well-spaced, deeply sunken pits, in the centre of which is a slight eminence—a structure which has almost the character of a granule surrounded by a scrobicule (see Pl. IV, fig. 2 *b*). On the surface which stands in the margin the punctations are fewer and more widely spaced on the upper half of the surface—that is to say, the half adjacent to the supero-marginal series (see Pl. IV, fig. 4 c). I have not found any pedicellariæ on these plates.

The adambulacral plates are broader than long, except on the outer part of the ray, and bear on the surface four or five ridges with granuliform eminences, upon which the spinelets constituting the adambulacral armature were originally borne. In one well-preserved specimen these small articulatory tubercles are seen to have each a small microscopic central puncture (see Pl. IV, fig. 3 c), but I am not certain whether this is always present.

The actinal interradial areas are small, and are covered with a comparatively small number of large pentagonal or tetragonal intermediate plates, which are arranged in series parallel to the ambulacral furrow, and form a compact tessellated pavement. The actinal intermediate plates are larger in relation to the size of the disk than in the species above described. The plates of the series adjacent to the adambulacral plates, and a few of the plates of the succeeding series within the angle towards the mouth, are larger than the others. The intermediate plates do not extend beyond the second, or at most a short distance along the margin of the third infero-marginal plate, counting from the median interradial line. The surface of the intermediate plates, excepting a border round the margin of the plate, is covered with large punctations, which are nearly confluent, and in some cases almost give the appearance of a coarse reticulate superficial ornamentation; the border round the margin of the plate above mentioned is marked with a concentric crenulation (see Pl. IV, fig. 3 b). Within the pits are more or less definite elevations. In other examples the reticulate character is less marked, and the margin of the pit is then prominently lipped, and the marginal crenulation is not so strongly shown (see Pl. IV, fig. 4 e).

In the marginal view of the type specimen the supero-marginal plates are seen to be higher than the infero-marginal plates, and that their height is greater than their length, whereas in the infero-marginal series of plates the height is less than the length (see Pl. IV, fig. 2 c).

In other examples the mouth-plates are preserved. These are rather small, triangular, and covered with rather large, irregular, tuberculose eminences for the attachment of the mouth-plate armature.

Variations.—Three examples of this species are figured on Plate IV. These present a number of minor differences, which will be readily noticed on referring to the figures.

The example which is shown in fig. 3 a has the marginal border of the infero-marginal plates rather less broad than in the type form, and it is especially remarkable for the peculiar retiform and crenulated ornamentation of the actinal intermediate plates already noticed. The disposition of the armature of the

PLATE I.

CALLIDERMA SMITHIÆ, *Forbes*, sp.   (P. 6.)

From the Lower Chalk.

FIG.

1 *a.* Actinal aspect; natural size.   (Coll. Brit. Mus.)

   b. Lateral view of the margin ; natural size.

   c. Infero-marginal plates ; magnified 2 diameters.

   d. Lateral view of the marginal plates ; magnified.

   e. Adambulacral plates ; magnified.

   f. Actinal intermediate plates ; magnified 3 diameters.

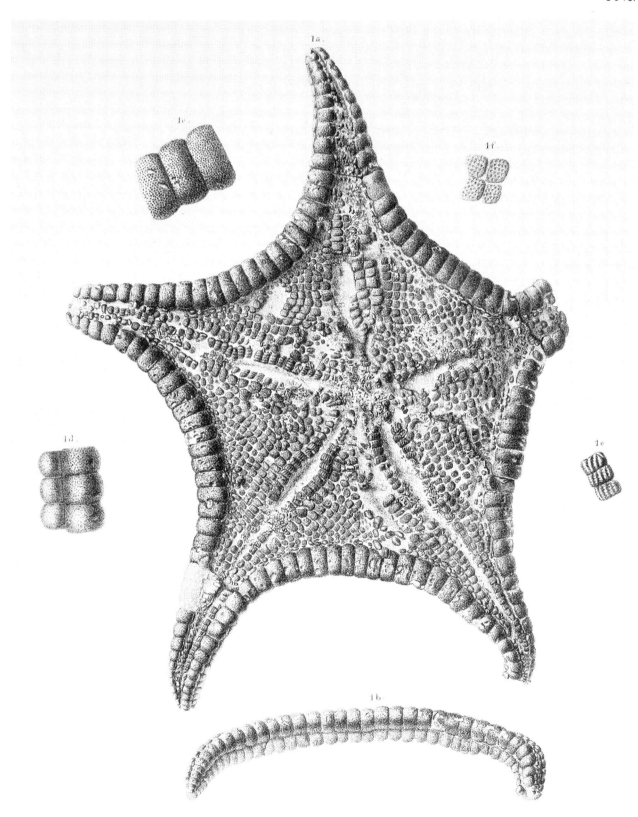

A.H.Searle del.et lith .

Hanhart imp.

PLATE II.

CALLIDERMA LATUM, *Forbes*, sp.   (P. 12.)

From the Lower Chalk.

FIG.

1 *a.* Actinal aspect; natural size.   (Coll. Brit. Mus.)

  b. Infero-marginal plates; magnified 2 diameters.

  c. Actinal intermediate plates; magnified 3 diameters.

  d. Adambulacral plates; magnified.

  e. A mouth-plate; magnified.

2 *a.* Actinal aspect of another example, with a portion of the actinal floor removed, showing the stellate bases of the abactinal plates or paxillæ; natural size. (Coll. Brit. Mus.)

  b. Actinal surface of part of the ray; magnified $1\frac{1}{2}$ diameters.

  c. Infero-marginal plates; magnified 3 diameters.

  d. Stellate bases of the abactinal plates or paxillæ; magnified 4 diameters.

Pl. 11.

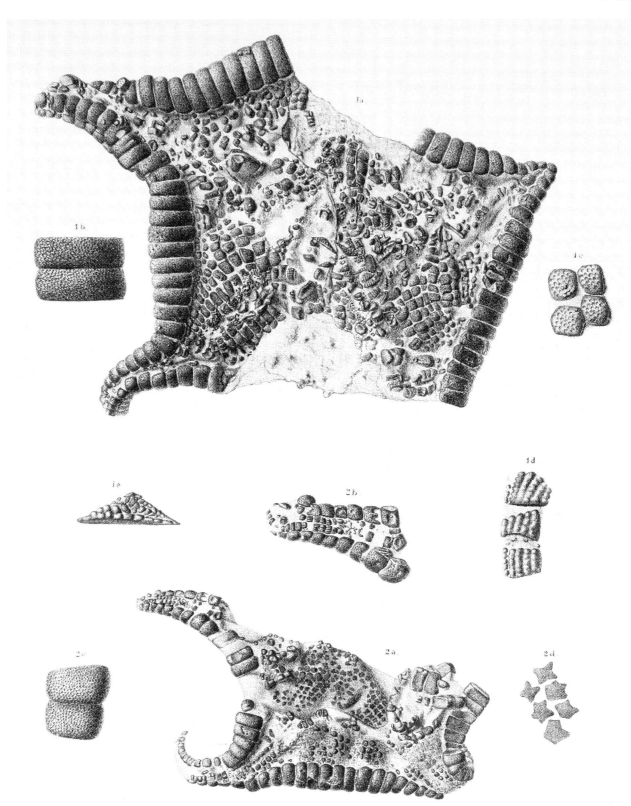

A.H.Searle del.et lith.

Hanhart imp.

PLATE III.

CALLIDERMA LATUM, *Forbes*, sp. (P. 12.)

From the Lower Chalk.

FIG.

1 *a*. Actinal aspect of an example from Amberley ; natural size.   (Coll. Brit. Mus.)

  b. Lateral view of the margin ; natural size.

  c. Lateral surfaces of the marginal plates ; magnified.

  d. An infero-marginal plate ; magnified.

  e. An actinal intermediate plate ; magnified.

2 *a*. Actinal aspect of an example from the Chalk Marl of Dover ; natural size. (Coll. Brit. Mus.)

  b. An adambulacral plate ; magnified.

3 *a*. Actinal aspect of an example from Washington, with a portion of the actinal floor removed, showing the stellate bases of the abactinal plates or paxillæ ; natural size.   (Coll. Brit. Mus.)

  b. Stellate bases of the abactinal plates or paxillæ ; magnified.

Pl. III.

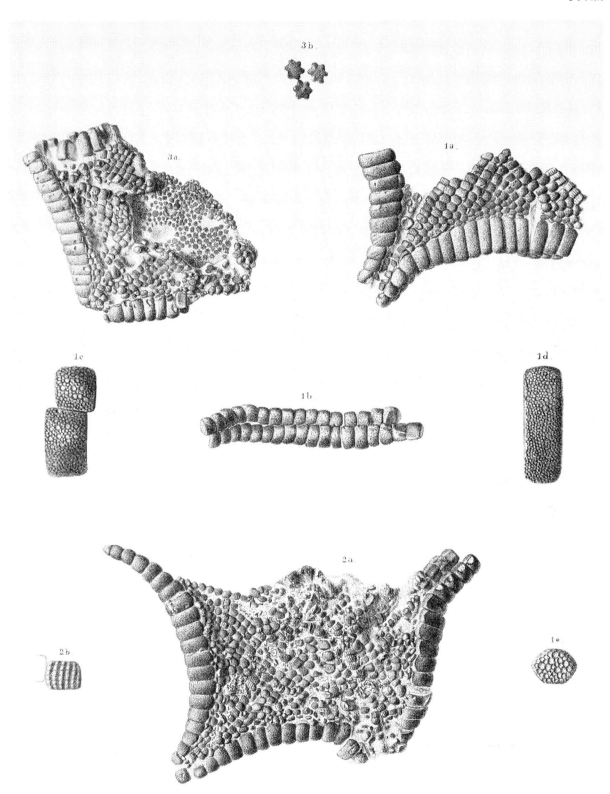

A.H.Searle del.et lith.

Hanhart imp.

PLATE IV.

PENTAGONASTER LUNATUS, *Woodward*, sp.   (P. 25.)

From the Upper Chalk.

Fig.

1 *a*. Actinal aspect of the type specimen; natural size.   (Coll. Norfolk and Norwich Mus.)

  b. An actinal intermediate plate; magnified.

  c. An infero-marginal plate; magnified.

PENTAGONASTER MEGALOPLAX, *Sladen*.   (P. 27.)

From the Lower Chalk.

2 *a*. Actinal aspect of the example figured by Forbes, under the name of *Goniaster* (*Astrogonium*) *lunatus*; natural size.   (Coll. Brit. Mus.)

  b. An infero-marginal plate; magnified 3 diameters.

  c. Lateral view of the margin; natural size.

3 *a*. Actinal aspect of another example; natural size.   (Coll. Brit. Mus.)

  b. An actinal intermediate plate; magnified 6 diameters.

  c. Adambulacral plates; magnified 6 diameters.

4 *a*. Actinal aspect of an example from the Upper Chalk of Bromley; natural size.   (Coll. Brit. Mus.)

  b. Lateral view of the margin; natural size.

  c. Lateral surface of an infero-marginal plate; magnified.

  d. An infero-marginal plate; magnified.

  e. An actinal intermediate plate; magnified 6 diameters.

Pl. IV.

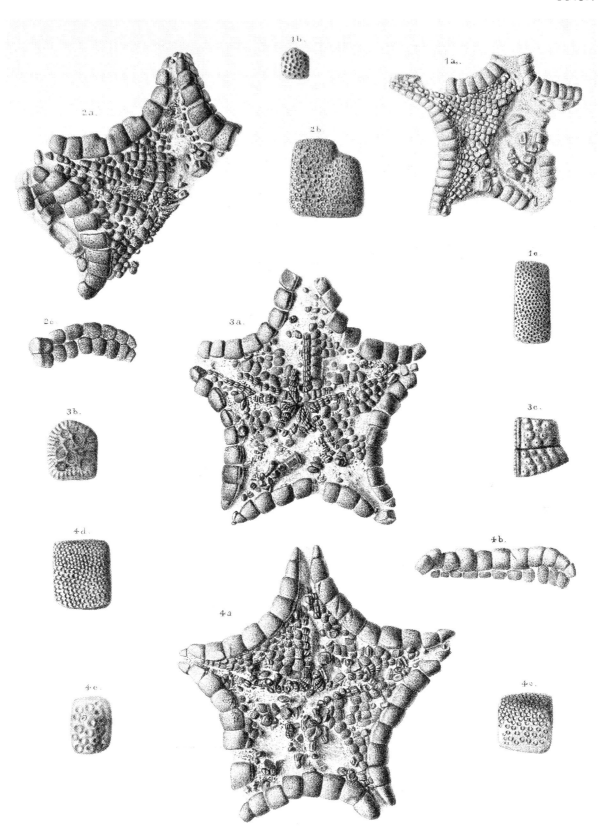

A.H.Searle del.et lith.

Hanhart imp.

PLATE V.

Tomidaster sulcatus, *Sladen.*

From the Grey Chalk.

Fig.

1 *a.* Actinal aspect; natural size. (Coll. Brit. Mus.)

  b. Actinal intermediate plates; magnified.

  c. Adambulacral plates; magnified 3 diameters.

Calliderma mosaicum, *Forbes,* sp. (P. 9.)

From the Lower Chalk.

2 *a.* Abactinal aspect; natural size. (Coll. Brit. Mus.)

  b. Lateral view of the margin; natural size.

  c. Abactinal plates; magnified.

  d. Supero-marginal plates; magnified.

  e. Madreporiform body and surrounding plates; magnified.

PI . V.

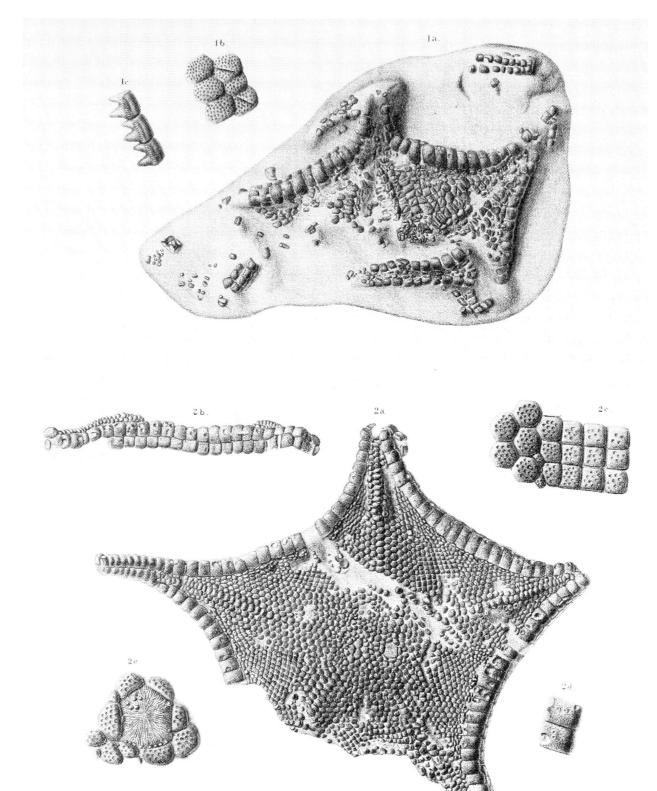

A.H.Searle del.et lith.

Hanhart imp.

PLATE VI.

Calliderma mosaicum, *Forbes*, sp.   (P. 9.)

From the Grey Chalk.

Fig.

1. Abactinal aspect of an example from which the abactinal plates have been
   removed; natural size.   (Coll. Brit. Mus.)
2 *a.* Abactinal aspect of another example from which the abactinal plates have
   been removed, showing the inner surface of some of the actinal intermediate
   plates and adambulacral plates; natural size.   (Coll. Brit. Mus.)
   b. Adambulacral plates, seen from within; magnified.
   c. Actinal intermediate plates, inner surface; magnified.

1

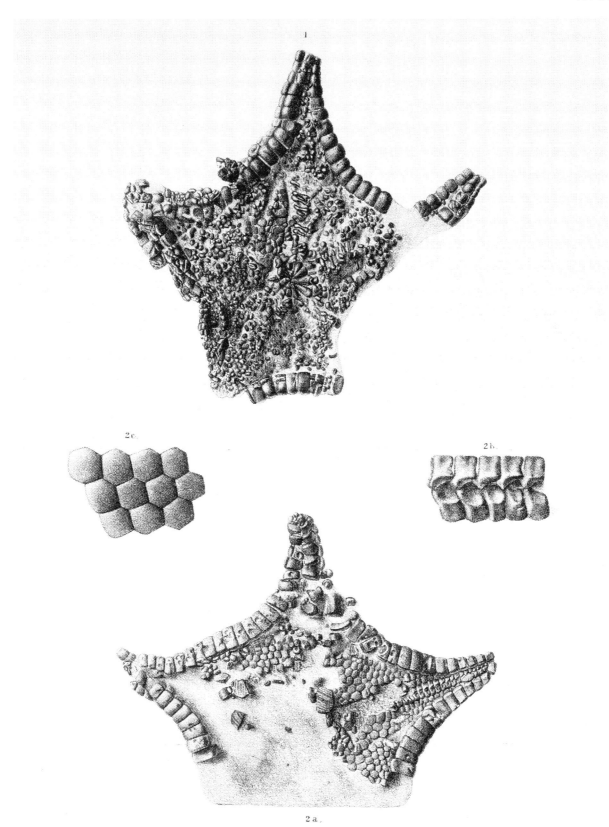

A.H.Searle del.et lith .

Hanhart imp.

PLATE VII.

(?) Nymphaster Coombii, *Forbes*, sp.   (P. 15.)

From the Lower Chalk.

Fig.

1 *a.* Actinal aspect; natural size.   (Coll. Brit. Mus.)

  b. Lateral view of the margin; natural size.

  c. Adambulacral plates; magnified.

  d. An infero-marginal plate; magnified.

  e. Actinal intermediate plates; magnified.

2 *a.* Abactinal aspect of an example from the Grey Chalk at Folkestone; natural size.   (Coll. Brit. Mus.)

  b. A supero-marginal plate; magnified.

3 *a.* Abactinal aspect of an example from the Lower Chalk of Glynde; natural size.   (Coll. Brit. Mus.)

  b. A supero-marginal plate; magnified.

Calliderma mosaicum, *Forbes*, sp.   (P. 9.)

From the Lower Chalk.

4 *a.* Abactinal aspect; natural size.   (Coll. Brit. Mus.)

  b. Lateral view of the margin; natural size.

  c. A supero-marginal plate; magnified.

Pl. VII.

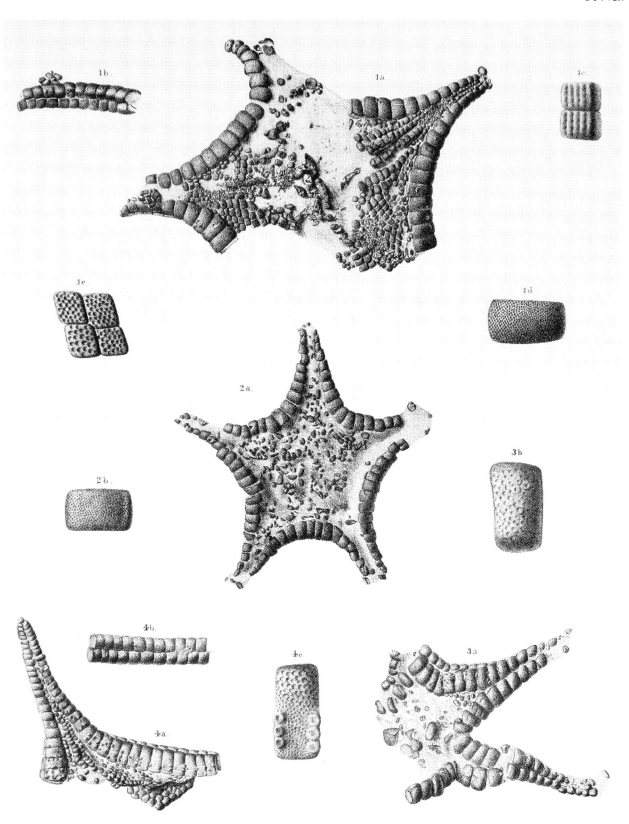

A.H.Searle del.et lith.

Hanhart imp.

PLATE VIII.

NYMPHASTER COOMBII, *Forbes*, sp.   (P. 15.)

From the Lower Chalk.

FIG.
1 *a.* Actinal aspect of the type specimen ; natural size.   (Coll. Brit. Mus.)
   b. An infero-marginal plate ; magnified 3 diameters.

CALLIDERMA SMITHIÆ, *Forbes*, sp.   (P. 6.)

From the Lower Chalk.

2 *a.* Profile view of a fragment of a ray ; natural size.   (Coll. Brit. Mus.)
   b. Actinal view of the same ; natural size.
   c. An adambulacral plate ; magnified 6 diameters.

NYMPHASTER OLIGOPLAX, *Sladen*.   (P. 19.)

From the Upper Chalk.

3 *a.* Abactinal aspect ; natural size.   (Coll. Brit. Mus.)
   b. A supero-marginal plate ; magnified.

NYMPHASTER MARGINATUS, *Sladen*.   (P. 18.)

From the Upper Chalk.

4 *a.* Abactinal aspect ; natural size.   (Coll. Brit. Mus.)
   b. A supero-marginal plate ; magnified 4 diameters.

Pl. VIII.

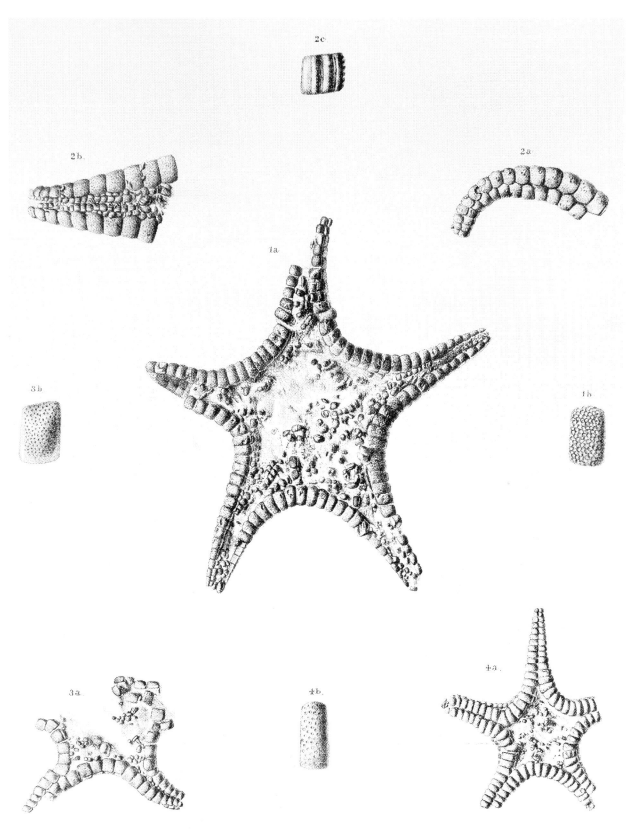

A.H.Searle del.et lith.

Hanhart imp.

THE

PALÆONTOGRAPHICAL SOCIETY.

INSTITUTED MDCCCXLVII.

VOLUME FOR 1893.

LONDON:

MDCCCXCIII.

A MONOGRAPH

ON THE

BRITISH FOSSIL

ECHINODERMATA

FROM

THE CRETACEOUS FORMATIONS.

VOLUME SECOND.
THE ASTEROIDEA.

BY

W. PERCY SLADEN, F.L.S., F.G.S., &c.,

SECRETARY OF THE LINNEAN SOCIETY.

PART SECOND.

PAGES 29—66; PLATES IX—XVI.

LONDON:

PRINTED FOR THE PALÆONTOGRAPHICAL SOCIETY.

1893.

PRINTED BY ADLARD AND SON,
BARTHOLOMEW CLOSE, E.C., AND 20, HANOVER SQUARE, W.

adambulacral plates is also noteworthy. This seems to indicate the former presence of a distinct furrow series of spinelets or granules much smaller than usual, followed by granules or spinelets borne on the outer part of the plate, more irregularly placed than in the other forms described, and articulated on punctured eminences.

The example which is represented in fig. 4 a also has a narrower marginal border of infero-marginal plates than the type. The punctation of the infero-marginal plates is smaller than in the type, and does not present the striking scrobiculate character noticed in that example. The markings are rather to be described as lipped pits, and some granules are still *in situ*. The actinal intermediate plates do not have the retiform and crenulate ornamentation shown in the plates belonging to the specimen figured in 3 a, but the margins of the punctations are strongly lipped. The supero-marginal plates are less regular and much less high than in the type specimen, but they are not perfectly preserved.

Dimensions.—In the type specimen (figured on Pl. IV, fig. 2 a) the major radius is about 41 mm., and the minor radius 26 mm. Breadth of a ray between the third and fourth infero-marginal plates, counting from the median interradial line, about 12 mm. or rather more. Thickness of the margin about 8 5 mm.

The specimen given in fig. 3 a has a major radius of about 39 mm. and a minor radius of 24 mm.

The specimen given in fig. 4 a has a major radius of about 41 mm. and a minor radius of 25·5 mm. Breadth of the ray between the third and fourth infero-marginal plates about 11 to 12 mm., or rather more.

Locality and Stratigraphical Position.—The type specimen, which is now preserved in the British Museum, is labelled from the " Lower Chalk " of " Sussex," but is stated by Forbes to have been obtained from the Upper Chalk. Other examples of the species have been collected from the Upper Chalk of Bromley, Sittingbourne, Purfleet, Gravesend, Sussex, and Wiltshire. Fine series are preserved in the British Museum and in the Museum of Practical Geology, Jermyn Street.

History.—The specimen which I have taken as the type of this species was originally referred by Forbes to the *Asterias lunatus* of Woodward, and was figured by him as that species in Dixon's ' Geology and Fossils of the Tertiary and Cretaceous Formations of Sussex ' (pl. xxiii, fig. 9). The same example is carefully represented on Pl. IV, fig. 2 a, of this memoir.

5

Remarks.—It will be at once seen on referring to the figures that Woodward's *Pentagonaster lunatus*, which is drawn on Pl. IV, fig. 1 *a*, of this Monograph, is a distinct species. The rays are more produced, and are narrower at the base. The infero-marginal plates are twice as numerous, the marginal border is less broad, and the plates are much shorter in proportion to their breadth. Their punctation is also different. The actinal intermediate plates are smaller in relation to the size of the actinal interradial areas, and their punctation is different from that which characterises *Pentagonaster megaloplax*. The armature of the adambulacral plates also appears to be more regular in its arrangement.

Under these circumstances I have no hesitation in considering the form under description a distinct species. I much regret having to impose a new name, as this form has for a long time been known under the specific name of *lunatus*; the course, however, seems unavoidable, as the actual type of the real *Pentagonaster lunatus* described by Woodward is in existence, and there can, in my opinion, be no question as to its being a different species.

Since the preceding sheet was printed off I have found several specimens in the British Museum which show the abactinal aspect of the disk. I have little hesitation in referring these examples to *Pentagonaster megaloplax*, and a drawing of one of them is given on Pl. XIII, fig. 1 *a*. The infero-marginal plates all show more or less distinctly the characteristic "scrobiculate" or areolated pits already described. A similar ornamentation also extends upon the supero-marginal plates, but is confined to the lateral wall which falls in the margin of the disk. The curvature which unites the abactinal and lateral areas of the plate is more or less abrupt, and the lateral wall of the disk is consequently vertical and not rounded, as a rule. The abactinal area of the supero-marginal plates is covered with small, uniform, granular eminences (see Pl. XIII, fig. 1 *b*). Two or three supero-marginal plates at the extremity of the ray meet the corresponding plates of the opposite side of the ray in the median radial line, and a rapid diminution in breadth occurs as they approach the extremity.

Genus—METOPASTER, *Sladen.*

[Μέτωπον = a cheek-piece.]

Body depressed and pentagonal or stellato-pentagonal in contour, the rays being produced to a very slight degree. Marginal plates covered with well-spaced uniform punctations, upon which granules were originally borne, and surrounded by a narrow depressed border with very minute and crowded punctations for the articulation of setæ. Supero-marginal plates ordinarily few in number, and form a broad border to the disk. Ultimate paired supero-marginal plates the largest of the series. Abactinal area covered with polygonal, and usually hexagonal, plates (some of which may have stellate bases), and upon the tabulæ are borne small, more or less co-ordinated granules. Infero-marginal plates more numerous than the supero-marginal series, and decreasing in size as they approach the extremity of the ray. Their ornamentation similar to that of the superior series. Actinal intermediate plates small, polygonal, covered with uniform granules. Armature of the adambulacral plates arranged in longitudinal lines. Small entrenched pedicellariæ may be present occasionally on the plates.

Metopaster differs from *Pentagonaster* by the large ultimate paired supero-marginal plates, by the comparatively small number of the supero-marginal plates, which are also fewer in number than the infero-marginal series, and by the character of the ornamentation of the marginal plates of both series.

The forms ranked under this genus were all classed by the late Professor Edward Forbes under *Goniodiscus*, which he considered to be a sub-genus of *Goniaster*. There is, however, no justification whatever in my opinion for regarding any of the Cretaceous starfishes hitherto described as belonging to either the genus *Goniaster* or *Goniodiscus*. The species which may be considered as the types of each of these genera are existing forms, and no Cretaceous forms agreeing in structural detail have, so far as I am aware, been discovered. It is also erroneous to rank *Goniodiscus* as a sub-genus of *Goniaster*. The two genera belong to different families; and I am in perfect accord with Professor Edmond Perrier as to the limitation of the two genera. His view appears to me to be perfectly logical, and to be the result of careful and impartial judgment. I also consider that the fossil forms under consideration are quite distinct from the recent genus *Astrogonium*, as limited by me elsewhere.[1]

1. METOPASTER PARKINSONI, *Forbes*, sp. Pl. IX, figs. 2 *a*—2 *c*; Pl. X, figs. 1 *a*— 5 *c*; Pl. XI, figs. 1 *a*—2 *c*; Pl. XII, figs. 1 *a*—1 *d*; Pl. XVI, figs. 2 *a*, 2 *b*.

PENTAGONASTER REGULARIS, *Parkinson*, 1811. Organic Remains, vol. iii, p. 3, pl. i, fig. 3 (*non* Linck).
TOSIA REGULARIS, *Morris*, 1843. Catalogue of British Fossils, p. 60.

[1] 'Zool. Chall. Exped.,' part li, "Report on the Asteroidea," 1889, p. 285.

Goniaster (Goniodiscus) Parkinsoni, *Forbes*, 1848. Memoirs of the Geological Survey of Great Britain, vol. ii, p. 472.

—   —   Rectilineus, *McCoy*, 1848. Ann. and Mag. Nat. Hist., ser. 2, vol. ii, p. 408.

—   —   Parkinsoni, *Forbes*, 1850. In Dixon's Geology and Fossils of the Tertiary and Cretaceous Formations of Sussex, London, 4to, p. 332, pl. xxi, figs. 10, 11; pl. xxii, figs. 5—7.

—   —   —   *Morris*, 1854. Catalogue of British Fossils, 2nd ed., p. 81.

—   —   Rectilineus, *McCoy*, 1854. Contrib. Brit. Pal., p. 55.

—   —   —   *Morris*, 1854. Catalogue of British Fossils, 2nd ed., p. 81.

Astrogonium Parkinsoni,   *Dujardin and Hupé*, 1862. Hist. Nat. Zooph. Échin. (Suites à Buffon), p. 399.

—   Rectilineum,   *Dujardin and Hupé*, 1862. Ibid., p. 400.

Goniaster (Goniodiscus) Parkinsoni, *Forbes*, 1878. In Dixon's Geology of Sussex (new edition, Jones), p. 365, pl. xxiv, figs. 10, 11; pl. xxv, figs. 5—7.

Pentagonaster rectilineus,   *Woods*, 1891. Catalogue of the Type Fossils in the Woodwardian Museum, Cambridge, p. 36.

Body of medium size. General form depressed. Abactinal surface flat, with a tendency, however, for the extremity of the rays to be slightly upturned; as found in the fossil state the area occupied by the abactinal plates is usually at a lower level than the marginal plates, which leads to the assumption that the abactinal floor had collapsed or fallen to a certain extent on the death of the animal. Actinal surface slightly convex. Marginal contour pentagonal, with slightly lunate sides, the curvature being often flattened at right angles to the median interradial line. The major radius measures about one-third more than the minor radius, and frequently less than one-third; the rays are consequently very feebly produced. Margin thick and well rounded.

The supero-marginal plates are four in number, counting from the median interradial line to the extremity, or eight from the tip of one ray to the tip of the adjacent ray, exclusive of the odd terminal or "ocular" plate in each case. They form a broad border to the abactinal area of the disk of uniform breadth throughout, which measures about 9 mm. at the median interradial line in an example whose minor radius measures 30 mm. Excepting the ultimate paired plates all the supero-marginal plates are of equal size, the breadth being about twice the length, the actual measurements in the specimen under notice being, length 4·75 mm. and breadth 9·5 mm. respectively, *i. e.* as 1 : 2. The abactinal surface of these plates is distinctly convex, with a slight depression along their margins of juncture, formed by a well-defined bevel along the sides and adcentral end. The general surface of the whole series is well rounded, the curvature being regular and uninterrupted between the adcentral margin of the plate and the margin in the lateral wall adjacent to the infero-marginal plates. The height of the plates as seen in the margin is a little greater than their length, and there is no diminution in height as the plates approach the extremity of the ray—in fact, the ultimate paired plate is not unfrequently higher than the other plates in consequence of a tendency to become gibbous on its abactinal surface. The whole superficies of the plates is covered with small, widely spaced, equidistant, uniform punctations, and there is a depressed border along the margin of the plate, varying slightly in breadth in different examples, covered with much smaller and closely crowded punctations, upon which much smaller granules than those upon the median area of the plate were originally borne. Traces of these granules may occasionally be found *in situ.*

The ultimate paired plate is larger than any of the other supero-marginal plates, and is of a different shape. It is subtriangular in form as seen from above, and one margin touches the corresponding plate of the adjacent side of the disk throughout, the junction coinciding with the median radial line of the disk. The length of this margin of the plate is subequal to or only slightly greater than the breadth of the preceding marginal plates. In small specimens the subequal measures are the rule, whilst in larger examples the plate becomes more elongate and produced in the direction of the prolongation of the ray. When viewed in the margin of the test the form of the ultimate plate strikingly resembles that of the carapace of some Coleoptera. The length of the plate in this aspect, measured from its outer extremity to the margin adjacent to the penultimate plate, is in small and medium sized specimens about once and a half the length of the other marginal plates, but in large examples it may be as much as, or even exceed, twice their length. The surface of the ultimate plate bears a similar ornamentation to that on the other supero-marginal plates.

The odd terminal plate is very small, appearing externally when denuded of granules like a truncate cylinder, having a fanciful resemblance to a cannon projecting from a porthole. This plate seems to be very rarely preserved *in situ* in the fossil state. In a remarkably good specimen belonging to the British Museum Collection (marked " E 2034 ") (see Pl. XVI, figs. 2 *a*, 2 *b*) each of the terminal plates preserved bears at its outer truncate extremity a single horizontally placed entrenched pedicellaria. Whether this regularly placed pedicellaria is always present on the odd terminal plate in this species I am unable to say.

The abactinal area of the disk within the boundary of the marginal plates is covered with small sub-regular plates or paxillar tabulæ, an hexagonal form predominating especially in the radial regions; and a small but distinct diminution in size takes place as the plates approach the margin of the disk. All the plates have their surface marked with minute, shallow, and closely placed punctations—the impressions of the attachment of the granules previously present. The primary basal plates are larger than any of the other abactinal plates, and they are well shown in several of the drawings illustrative of this species (see Pl. X, figs. 1, 2 a). Occasional plates bear small entrenched pedicellariæ, the normal form consisting of a small, central, lipped foramen with a lateral trench on each side. It frequently happens, however, that the organ exhibits a more complex development, and assumes a stellate form in consequence of the presence of additional trenches—five or six being not an unusual number—which radiate from the central foramen; the whole being placed on a small hemispherical elevation, and producing an appearance shown on Pl. X, fig. 2 d.

The madreporiform body is large and subtriangular in outline; and its surface is sculptured by very fine striations which radiate from the centre to the margin, with more or less wavy lines here and there (see Pl. X, fig. 2 c). The margin of the plate is surrounded by three large plates, one on the adcentral side of the madreporite towards which it presents a straight suture; the other two plates are on the remaining sides of the triangular body, and they have a concave curve directed towards the madreporite to correspond with the convexity of its sides. The position of the madreporite is nearer the centre of the disk than the margin.

The infero-marginal plates are seven in number, counting from the median interradial line to the extremity—that is to say, there are fourteen for the whole side of the disk, as against eight in the supero-marginal series. The length of the three innermost plates on each side of the median interradial line is equal to that of the superior series, but there are four infero-marginal plates corresponding to the large ultimate supero-marginal. As seen in the lateral wall of the disk the height of the infero-marginal plates is less than that of the supero-marginal series. The breadth of these plates adjacent to the median interradial line on the actinal surface is 7·5 mm. in an example whose major radius measures 36·5 mm. and the minor radius 27·5 mm. The breadth of the marginal border rapidly diminishes towards the extremity of the ray. The surface of the plates is ornamented in a precisely similar manner to that of the supero-marginal plates. A narrow border of smaller granulation is also present round the whole margin of the plate, similar in all respects to that already described in the case of the superior series of plates.

The adambulacral plates are small, about or nearly twice as broad as long, and their surface is traversed by four or five ridges running parallel to the ambulacral furrow, with punctures upon which the spinelets composing the armature were articulated. There were four or five spinelets in each lineal series. The spinelets are short, their length being about equal to the length of the plate, stumpy, compressed, slightly tapering and rounded at the extremity; and all appear to have been uniform.

The actinal intermediate plates are rather large for the genus; those adjacent to the adambulacral plates are pentagonal in form, but elsewhere they are subhexagonal, or perhaps more correctly polygonal. The plates are very large on the inner portion of the area, but diminish greatly in size at the outer margin of the disk adjacent to the marginal plates. The surface of the intermediate plates is entirely covered with small, equidistant punctations, upon which a uniform close granulation was previously attached. Remains of this granulation are still occasionally to be seen *in situ* on plates here and there in the example under notice. Entrenched pedicellariæ similar to those above described on the plates of the abactinal surface occur on a number of the plates in the series adjacent to the adambulacral plates, but the organ does not appear to diverge, or only very rarely, from the normal form of a central foramen and two lateral trenches.

Dimensions.—In the specimen figured on Pl. X, fig. 2 *a*, the major radius is 38 mm., and the minor radius 30 mm. Other examples have the following approximate measurements : $R = 35$ mm., $r = 27$ mm.; $R = 43$ mm., $r = 34$ mm. The diameter of the disk ($R + r$) in well-grown tests ranges, therefore, from 60 mm. to 80 mm. The thickness of the margin is about $11\cdot75$ mm.

Locality and Stratigraphical Position.—All the examples figured in this Monograph were obtained from the Upper Chalk, near Bromley. The species is a characteristic Upper Chalk fossil in the south of England, and has been found in beds of that age at Brighton, Charlton, Gravesend, Kent, and other localities. It is stated by Forbes to occur in the Lower Chalk of Sussex, but I have not seen any examples from that horizon.

History.—A fossil starfish which has been generally considered to be this species was figured by Parkinson in his ' Organic Remains of a Former World,' vol. iii, pl. i, fig. 3, but it was referred by that author to the *Pentagonaster regularis* of Linck. The last named has, however, been supposed to be a recent species, but the type has unfortunately been lost, and the form has not subsequently been recognised definitely. Apart from this the fossil starfishes now under notice are certainly distinct from the form indicated by Linck's figure, and this view was taken by Forbes, who named the species after Parkinson in his memoir ' On the Asteriadæ found fossil in British Strata,' and figures of the fossils so named by him were given in Dixon's Geology and Fossils of the Tertiary and Cretaceous Formations of Sussex,' London, 1850. One, if not more, of the specimens delineated in that work is now preserved in the British Museum.

Variations.—In some examples the breadth of the border formed by the supero-marginal plates on the abactinal surface is greater in relation to the disk-area than in others, and this variety was noted by Forbes ('Mem. Geol. Surv.,' vol. ii, p. 472). I have not been able to establish the relation of this modification with any other permanent morphological character, nor to associate it with any special locality or stratum, and I am therefore led to consider, for the present at least, that the variation in question is one affecting individual examples of the species independently of other structural modifications which would warrant recognition by name.

Two other variations are to be noted in this species which are superficially much more striking, and either of them would, if only isolated examples were known, lend a strong temptation to the separation of their possessor from the normal form of the species. One of the variations in question affects the large ultimate paired supero-marginal plates. On comparing the examples drawn on Pl. X, fig. 1 and fig. 2 a, with fig. 1 a, Pl. XII, and fig. 1 a, Pl. XI, it will be seen that the ultimate plates are relatively much larger than the adjacent supero-marginal plates and are more produced at the extremity; whilst in the specimen delineated in fig. 2 a, Pl. XI, this modification is carried to such an extent that at first sight it would appear scarcely possible to believe that this fossil belongs to the same species as, for example, fig. 2 a, Pl. X. I have, however, been unable to find any other constant variation from what has been considered the typical form of *Metopaster Parkinsoni* associated with this modification in the size and shape of the ultimate plates; and as the most complete gradation between the two extremes may be traced in the splendid series of specimens now preserved in the British Museum, all obtained from the same locality and the same horizon, no reasonable doubt can be entertained that the variation in question is of a comparatively trivial character, affecting the individual independently, and that it is not stamped by correlation with other structural modification with sufficient importance to justify the forms being separated from the species, or even a name being given to the variety. All the examples referred to in the foregoing remarks and figured in the plates accompanying this Memoir are from the Upper Chalk, and were obtained from the same locality near Bromley.

The fine specimen with large and greatly produced ultimate plates drawn on Pl. XI, fig. 2 a, is also an example of the second variation in the structure of this species, to which I have alluded. This manifests itself in the presence of an additional supero-marginal plate. It will be seen that there are nine supero-marginal plates, exclusive of the terminals (or so-called oculars) on each side of the pentagonal disk. At first sight the presence of this odd intermediate marginal plate in association with the strikingly modified ultimate plates found in this example would appear sufficient to indicate a well-marked variety, if not actually a distinct species. A careful examination of the collection in the British Museum shows, however, that this assumption is untenable, for in another example (which bears the Museum number " 46,765 "), which also possesses an odd intermediate marginal plate on two of its sides, the ultimate plates are in no way specially abnormal in their form or size. The presence of the additional supero-marginal plate would in this case, therefore, appear to be only an occasional variation, and, so far as I am able to detect, occurring independently of other structural modifications. I have observed no differences either in the proportions or the ornamentation of the intermediate marginal plates worthy of remark. This view is strengthened, if not absolutely confirmed, by the presence in another specimen (British Museum number " 46,796 "), drawn on Pl. XII, fig. 1 a, of nine supero-marginal plates on two sides only of the disk, the remaining three sides having the normal number of eight marginal plates. In this example the ultimate plates are distinctly larger and more produced than in the truly normal forms of the species, but it will be noticed that their development is unequal, as is also the case in No. " 46,765 ; " in other words, in three of the rays one of the ultimate plates is smaller than the corresponding ultimate plate to which it is adjacent. The two sides which have an additional marginal plate are also the sides which have one of the smaller ultimate plates, and the inference naturally follows that the additional plate is to balance or compensate for the smaller size of the ultimate plate ; and I am inclined to think that the additional plate is not in this case a true *odd interradial* marginal plate at all, such as occurs in *Gnathaster*, Sladen, but that it is only a supplementary plate formed because its primitive or embryonic rudiment has not been included in the series merged together during development to form the structurally compound ultimate plate. That these ultimate plates are compound, or formed by the union of several embryonic plates, there is in my opinion little doubt when regard is had to the embryonic history of the ultimate plate and the associated infero-marginal plates.

2. Metopaster Mantelli, *Forbes*, sp.   Pl. XIII, figs. 2 *a*—4 *b*.

Pentagonaster semilunatus,	*Parkinson*, 1811. Organic Remains of a Former World, vol. iii, p. 3, pl. i, fig. 1 (*non* Linck).
? Goniaster semilunata,	*Mantell*, 1844. The Medals of Creation, London, vol. i, p. 338, lign. 75 (*non* Linck, ? *nec* Parkinson).
Goniaster (Goniodiscus) Mantelli,	*Forbes*, 1848. Memoirs of the Geological Survey of Great Britain, vol. ii, p. 472.
— — —	*Forbes*, 1850. In Dixon's Geology and Fossils of the Tertiary and Cretaceous Formations of Sussex, London, 4to, p. 332, pl. xxiii, figs. 11, 12.
— — —	*Morris*, 1854. Catalogue of British Fossils, 2nd ed., p. 81.
? — Mantelli,	*Mantell*, 1854. The Medals of Creation, London, 2nd ed., vol. i, p. 307, lign. 98.
Astrogonium Mantelli,	*Dujardin and Hupé*, 1862. Hist. Nat. Zooph. Échin. (Suites à Buffon), p. 399.
Goniaster (Goniodiscus) Mantelli,	*Forbes*, 1878. In Dixon's Geology of Sussex (new edition, Jones), p. 366, pl. xxvi, figs. 11, 12.

Body of medium or rather small size. General form depressed. Abactinal surface flat, with the area circumscribed by the supero-marginal plates plain and rather sunken below the level of the median convexity of the border formed by these plates. Actinal surface plain, or may be a little convex in consequence of a tendency to a slight upturning of the rays in some examples. Marginal contour pentagonal, with sides faintly lunate, or in small examples they may be almost straight. The major radius measures scarcely one-third more than the minor radius. Margin rather thin, but well rounded.

The supero-marginal plates are four in number, counting from the median interradial line to the extremity, or eight from the tip of one ray to the tip of the adjacent ray, exclusive of the odd terminal or "ocular" plate in each case. In one example, which is figured by Forbes in Dixon's 'Geology of Sussex,' there would appear to be only seven plates on the entire side, but only one radial angle is preserved intact, and much displacement of plates has occurred. The circumstance is in any case of comparatively trifling importance. The supero-marginal plates form a conspicuous and moderately broad border to the abactinal area of the disk, of uniform breadth throughout, and measuring about 6 mm. at the median interradial line in an example whose minor radius measures about 25 mm.

Excepting the ultimate paired plate all the supero-marginal plates are nearly subequal in size; the plates, however, adjacent to the median interradial line are slightly longer in relation to their breadth than those adjacent to the ultimate paired plate. The plates adjacent to the median interradial line have the appearance of being nearly square in outline as seen from above, the actual dimensions in the specimen under notice, whose side measures 38 mm., being length 5 mm., and breadth 6 mm. In a smaller example, with a side measurement of 29·5 mm., the corresponding plate is 3·5 mm. long and 4·5 mm. broad. In Forbes's type the measurements are, length 3·75 mm., and breadth 5 mm.

The abactinal surface of the supero-marginal plates is slightly tumid, and the general surface of the whole series forms a well-rounded regular curve from the adcentral margin to the margin in the lateral wall adjacent to the infero-marginal plates. The height of the plates as seen in the margin is less than their length, the actual measurement being about 3 mm. Their abactinal contour is distinctly convex, but not gibbous. The whole superficies of the plates is covered with small, widely spaced, equidistant, uniform punctations. In some examples which have been subjected to much weathering the punctations are almost obliterated, as in the case of the fine specimen shown on Pl. XIII, fig. 2 a. A narrow depressed border surrounds the margin of the plate, bearing very small, closely crowded punctations, those adjacent to the main or median area being in serial arrangement. Occasionally a small entrenched pedicellaria may be detected on the median area of the plate (see Pl. XIII, fig. 2 b).

The ultimate paired plate is small and triangular in outline as seen from above, and one margin touches the corresponding plate of the adjacent side of the disk throughout, the junction coinciding with the median radial line of the disk. The length of the plate—that is to say, of the side of the plate which falls in the margin of the disk—is a trifle greater than the length of the largest supero-marginal plate, measuring in the example under notice nearly 5·5 mm., whereas the breadth of the plate, or measurement of the side adjacent to the penultimate supero-marginal plate, is not more than 5 mm. Near the outer or apical extremity of this plate when seen from above there is frequently a more or less strongly developed tendency to gibbosity present.

The abactinal area of the disk within the boundary of the marginal plates is covered with small subregular hexagonal plates or paxillar tabulæ, which are, however, comparatively large for the size of the species. The primary apical plates are large and conspicuous, and all the plates in the central area of the disk and in the interradial areas are considerably larger than elsewhere; all diminish in size as they approach the margin of the disk. All the plates have their surface covered with a small, uniform, closely placed granulation, in which an indefinite subcircular arrangement in relation to the centre of the plate is discernible (see Pl. XIII, fig. 2 d).

The madreporiform body is small and subtriangular in outline; its surface is sculptured by fine striations which radiate from the centre to the margin. The madreporite is usually enclosed by three plates, but four may be present in consequence of the division or retarded development of one of them, as is the case in the example figured on Pl. XIII, fig. 2 c. The position of the madreporite is rather nearer the centre of the disk than the margin.

The infero-marginal plates are at least six in number, counting from the median interradial line to the extremity,—that is to say, there are twelve (or perhaps more) for the whole side of the disk, as against eight in the supero-marginal series. The length of the three innermost plates on each side of the median interradial line is equal to that of the superior series, but there are three infero-marginal plates corresponding to the ultimate paired supero-marginal. As seen in the lateral wall of the disk, the height of the infero-marginal plates is greater than that of the supero-marginal series. The breadth of these plates, adjacent to the median interradial line on the actinal surface, is 5 mm. in an example whose minor radius measures about 15 mm. The breadth of the marginal border appears to be well maintained till near the extremity. The surface of the plates is ornamented in a similar manner to that of the supero-marginal plates, excepting that the punctations on the main area are rather more numerous, and that the finely punctate depressed border round the margin of the plate is much broader than in the plates of the superior series; the border is broader on the adcentral margin of the plate than elsewhere. (Compare figs. 3 b and 4 b on Pl. XIII.)

The adambulacral plates appear to be comparatively small, but their preservation in the examples examined is not sufficiently good to permit of description.

The actinal intermediate plates, which are small and hexagonal, are covered with small, closely crowded, uniform granules. All are much displaced in the specimens under notice.

Dimensions.—In the example figured on Pl. XIII, fig. 2 a, the major radius is about 32 mm., and the minor radius about 25 mm. The length of the side is 38 mm. The thickness of the margin is about 8 mm. In Forbes's type (Pl. XIII, fig. 3 a) the major radius is about 20 mm., and the minor radius about 15·5 mm.;

the length of the side about 25·5 mm. In a very finely preserved cast from the Upper Cretaceous beds of Haldon the length of the side is 29·5 mm.

Locality and Stratigraphical Position.—All the examples of this form with which I am acquainted are from the Upper Chalk. One of Forbes's type is from Gravesend, but the locality of the other is not recorded. The large example figured on Pl. XIII, fig. 2 *a*, is from the Upper Chalk near Bromley.

An extremely well-preserved cast in flint from the Upper Cretaceous beds of Haldon (Devonshire) is in the collection of the Albert Museum, Exeter; and a cast of this example may be seen in the Museum of the Geological Survey, Jermyn Street.

History.—This form appears to have been first recognised by Parkinson, who erroneously referred it to the *Pentagonaster semilunatus* of Linck. The latter is a well-known recent species, and quite distinct from the fossil under consideration. Mantell, following Parkinson's determination, referred to the form under the name of *Goniaster semilunata.* Forbes was the first to indicate that these views of his predecessors were incorrect, and diagnosed the species in his memoir ' On the Asteriadæ found fossil in British Strata ' under the name of *Goniaster (Goniodiscus) Mantelli;* and figures of two examples were subsequently given in Dixon's ' Geology and Fossils of the Tertiary and Cretaceous Formations of Sussex,' London, 1850. Both these specimens are now preserved in the British Museum. Careful drawings of each fossil are given on Pl. XIII, figs. 3 *a* and 4 *a*.

Remarks.—It is not without hesitation that I maintain this species of Forbes's as independent from *Metopaster Parkinsoni.* For the present, however, I consider it to be distinguished by the smaller size, the comparative squareness of the supero-marginal plates, the small size of the ultimate paired plates, as well as by the character of the ornamentation of the supero-marginal plates and of the abactinal plates. Whether a more extensive series of specimens will break down or uphold these distinctions I do not feel prepared to say. It is undoubted that the two forms are very nearly allied.

I feel considerable doubt as to whether one of Forbes's types—that shown in Pl. XIII, fig. 4 *a*—really belongs to the same species as the examples illustrated in figs. 2 *a* and 3 *a* on the same plate.

3. Metopaster Bowerbankii, *Forbes*, sp.   Pl. XV, figs. 2 *a*—2 *d*; Pl. XVI, figs. 1 *a*—1 *d*.

Goniaster (Goniodiscus) Bowerbankii, *Forbes*, 1848.   Memoirs of the Geological Survey of Great Britain, vol. ii, p. 473.

—   —   —   *Forbes*, 1850.   In Dixon's Geology and Fossils of the Tertiary and Cretaceous Formations of Sussex, London, 4to, p. 333, pl. xxii, fig. 4.

—   —   —   *Morris*, 1854.   Catalogue of British Fossils, 2nd ed., p. 81.

Astrogonium Bowerbankii,   *Dujardin and Hupé*, 1862.   Hist. Nat. Zooph. Échin. (Suites à Buffon), p. 399.

Goniaster (Goniodiscus) Bowerbankii, *Forbes*, 1878.   In Dixon's Geology of Sussex (new edition, Jones), p. 366, pl. xxv, fig. 4.

Body of medium or rather large size. General form depressed. Abactinal surface flat; actinal surface also flat, or with a slight tendency to become convex. Marginal contour pentagonal, with a very small amount of lunation in the sides. Margin thick and well rounded.

The supero-marginal plates are five in number counting from the median interradial line to the extremity, or ten from the tip of one ray to the tip of the adjacent ray, exclusive of the odd terminal or "ocular" plate in each case. They form a broad border to the abactinal area of the disk of uniform breadth throughout, which measures about 8·5 mm. at the median interradial line in the type-specimen described by Forbes from the collection of the late Dr. Bowerbank. Excepting the ultimate paired plate all the supero-marginal plates are of equal size. They are short and broad, the breadth being nearly twice and a half the length, the actual measurements in the type being length 3·5 mm. or a trifle more, and breadth 8·5 mm., in which case the dimensions are in the proportion of 7 : 17. The abactinal surface of the plates is slightly convex along the direction of the breadth, sufficient to define each plate distinctly. The general surface of the whole series is gently arched towards the margin adjacent to the infero-marginal plates. The height of the plates as seen in the margin is about 6 mm., and there is scarcely any or only the slightest diminution in height as the plates approach the extremity of the ray, and the ultimate paired plate is not prominent or gibbous abactinally. The whole superficies of the supero-marginal plates is covered with small, widely spaced, equidistant, uniform punctations, and along the entire margin of the plate is a very narrow and deeply depressed border of fairly uniform breadth, covered with much smaller and closely crowded punctations, upon which much smaller granules than those which occupied the central area of the plate were originally borne. One or two or even three small entrenched pedicellariæ may be present on the central area of a plate, irregularly disposed.

The ultimate paired plate is triangular in form as seen from above. The margin or side which represents its length and coincides with the margin of the disk measures a little more than once and a half or nearly twice the length of the adjacent supero-marginal plates; and the margin representing the breadth of the plate which abuts on the penultimate supero-marginal plate is shorter than the margin of the latter plate. The remaining side of the plate which touches the corresponding ultimate plate of the adjacent side of the disk, and falls in the median radial line of the disk, is subequal to or even slightly shorter than the breadth of the preceding marginal plates. The ultimate plate in this species is not elongated, and no prolongation beyond the normal pentagonal contour of the disk occurs in the extension of the median radial line. The surface of the ultimate plate is covered with punctations and margined with a finer and closely crowded series precisely similar to those on the other supero-marginal plates.

The abactinal area of the disk within the boundary of the marginal plates is covered with small, subregular, hexagonal plates or paxillar tabulæ, which have their surface marked with minute, low, subhemispherical, and closely placed miliary granulations, which do not, however, extend quite to the margins of the plates. A number of the plates bear in the centre a rather large entrenched pedicellaria, consisting of a central foramen and normally two lateral fossæ, and there is usually a circular series of coalesced granules in which the fossæ are included, which imparts a very characteristic appearance to the organ in this species (see Pl. XVI, fig. 1 d).

The following description of the characters of the actinal area of the disk of this species is taken from an example preserved in the Museum of Practical Geology, Jermyn Street, and figures of which are given on Pl. XV, figs. 2 a—2 d.

The infero-marginal plates are eight in number counting from the median interradial line to the extremity,—that is to say, there are sixteen for the whole side of the disk, as against ten in the supero-marginal series. The length of the four innermost plates on each side of the median interradial line is slightly greater than that of the corresponding plates of the supero-marginal series; and there are four infero-marginal plates much smaller than those preceding, corresponding to the ultimate supero-marginal plate. As seen in the lateral wall of the disk the height of the infero-marginal plates is slightly greater than that of the supero-marginal series. The breadth of these plates adjacent to the median interradial line on the actinal surface is 6 mm., or even a little more in an example whose major radius measures 37 mm. and the minor radius 27 mm. The breadth of the marginal border rapidly diminishes towards the extremity of the ray. The surface of the infero-marginal plates is ornamented with extremely small and closely crowded punctations, upon which traces of a minute, closely crowded, and uniform granulation are preserved here and there.

The adambulacral plates are small, about twice as broad as long, and their surface is traversed by about three ridges, bearing punctures, running parallel to the ambulacral furrow, upon which the spinelets composing the adambulacral armature were articulated. There were about five or six spinelets in each lineal series.

The mouth-plates are regularly triangular, about twice and a half as long as broad, and the two adjacent plates which constitute a pair form together a regular rhomboid or lozenge-shaped figure. Their surface is covered with small, crowded, rather coarse, irregularly disposed tubercles or granules (see Pl. XV, fig. 2 d).

The actinal intermediate plates are fairly large, and there is a distinct diminution in size towards the outer margin of the disk adjacent to the marginal plates. The plates adjacent to the adambulacral plates are pentagonal, but a subhexagonal or polygonal form elsewhere is the rule, with comparatively little irregularity. The surface of the actinal intermediate plates is entirely covered with very small equidistant punctations, upon which a uniform close granulation was previously borne. Occasional small excavate pedicellariæ are present here and there, the lateral fossæ being slightly curved.

Dimensions.—The type specimen is unfortunately fragmentary, and the radial dimensions cannot be given. The length of one side of the disk, measured from the tip of one ray to the tip of the adjacent ray, is about 41 mm. ; the breadth of the supero-marginal plates adjacent to the median interradial line is 9 mm. ; and the thickness of the margin is about 12 mm.

In the fine example preserved in the Museum of Practical Geology, Jermyn Street, figured on Pl. XV, fig. 2 a, the major radius measures 37 mm., and the minor radius 27 mm.; the length of one side of the disk measured about 42 mm., or probably rather more when complete ; the breadth of the infero-marginal plates adjacent to the median interradial line is 6 mm. ; and the thickness of the margin is 8·5 mm.

Locality and Stratigraphical Position.—The type specimen is stated by Forbes to have been obtained from the Upper Chalk of Kent, but no indication of locality is now preserved on the label. The example belonging to the Museum of Practical Geology is from the Upper Chalk of Gravesend.

History.—The type of this species, which was described by Forbes in his memoir ' On the Asteriadæ found fossil in British Strata ' (' Mem. Geol. Surv.,' vol. ii, p. 473, 1848), originally formed part of Dr. Bowerbank's Collection, and was first figured by Forbes in Dixon's ' Geology and Fossils of the Tertiary and Cretaceous Formations of Sussex,' London, 1850, pl. xxii, fig. 4. That specimen is now preserved in the British Museum, where it bears the register number " E 2578." An accurate drawing of the type is given on Pl. XVI, fig. 1 *a*, of the present work.

4. METOPASTER ZONATUS, *Sladen.* Pl. XII, figs. 2 *a*—2 *c*.

Body of small or medium size. General form depressed. Abactinal surface in the fossil condition, as at present known, essentially concave in consequence of the conspicuously upturned extremities of the rays. Actinal surface conformably convex. Marginal contour pentagonal, with the radial angles slightly produced and obtusely rounded, and the sides distinctly lunate. The major radius measures nearly one-half more than the minor radius, or in the proportion of $3:2$ approximately; the actual dimensions in the example under notice being $R = 27$ mm., $r = 19$ mm. approximately. Margin very thick in relation to the size of the disk, and regularly rounded.

The supero-marginal plates are four in number, counting from the median interradial line to the extremity, or eight from the tip of one ray to the tip of the adjacent ray, exclusive of the odd terminal or " ocular " plate in each case. They form a broad border to the abactinal surface of the disk, of uniform breadth throughout, which measures about 8 mm. at the median interradial line, in an example whose minor radius measures 19 mm. Excepting the ultimate paired plate all the supero-marginal plates are subequal, the breadth being nearly three times the length, the actual measurements in the example under notice being length 2 75—3 mm., and breadth about 8 mm. The abactinal surface of the plates is distinctly convex along the median line of breadth, by which means each plate is conspicuously defined. The general surface of the whole supero-marginal series between the two ultimate plates is regularly rounded, and forms an uninter-

7

rupted regular curve between the adcentral margin of the plate and the margin in the lateral wall adjacent to the infero-marginal plates. The height of the plates as seen in the lateral view of the disk is greater than their length, and there is an apparent increase in height as the plates approach the extremity of the ray, the ultimate plate being still higher and distinctly tumid (see Pl. XII, fig. 2 *b*). The whole superficies of the plates is covered with small, widely spaced, equidistant, uniform punctations; and there is a narrow depressed border surrounding the margin of the plate with much smaller and closely crowded punctations, upon which minute miliary granules were previously borne.

The ultimate paired plate is much larger than any of the other supero-marginal plates, its length as measured on the outer margin being more than twice the length of the other supero-marginal plates. Its breadth is about equal to that of the adjacent supero-marginal plate. It is subtriangular in form as seen from above, and the line of junction with the corresponding companion plate of the adjacent side of the disk is complete throughout, and coincides with the median radial line. The convexity of the plate falls in a line parallel and adjacent to this margin of the plate, and its height is greatest there. From this convexity the surface slopes gradually and regularly in conformity to the curve of the superficies of the other supero-marginal plates. The actual dimensions of the ultimate plate in the example under notice are, length 7 mm., breadth 7 mm.; greatest height as seen in the marginal view, about 8·5 mm. The surface of the ultimate plate is marked with a precisely similar ornamentation to that on the other supero-marginal plates.

The odd terminal plate is very small, and though only traces are present in the type it is well preserved in other examples. It is prominent, cylindrical, and abruptly truncate, resembling in all respects the form described in *Metopaster Parkinsoni*.

The madreporiform body, which is only partially exposed in the example under notice, is apparently subtriangular in outline, and is marked with very fine centrifugally radiating striations (see Pl. XII, fig. 2 *c*).

The infero-marginal plates are seven in number, counting from the median interradial line to the extremity,—that is to say, there are fourteen for the whole side of the disk, as against eight in the supero-marginal series. The length of the three innermost plates on each side of the median interradial line is subequal to that of the superior series, but their height as seen in the marginal view of the disk is much greater, the thickness of the whole margin, *i. e.* both series of plates together, being 10 mm. in an example whose minor radius is 19 mm. Four plates underlie the superior ultimate plate, the last two being very small and triangular in form. All the four are adjacent to the adambulacral series of plates. The ornamentation or surface-marking of the infero-marginal plates is precisely similar to that of the supero-marginal series.

Dimensions.—In the example figured on Pl. XII, fig. 2 *a*, the major radius is 27 mm. and the minor radius about 19 mm., the thickness of the margin 10 mm.

Locality and Stratigraphical Position.—The figured example is from the Upper Chalk near Bromley. A number of other examples of this species are in the collection of the British Museum, all from the Upper Chalk, and most of them from the same locality, the locality of the remainder being not recorded.

Remarks.—I was at first somewhat in doubt as to whether to rank this form as a variety of *Metopaster Parkinsoni*, or as a distinct species. I have taken the latter course. Although there are many points of resemblance in general structure, as well as in various details when considered independently, the facies of this form is so distinct from that of *Metopaster Parkinsoni*, and is so readily recognisable, that there seems to me full justification for considering *Metopaster zonatus* specifically distinct. The great breadth of the supero-marginal plates as compared with their length, the form and character of the ultimate plates, the great thickness of the margin, and the relative proportions of the marginal plates, as well as the general habit of the abactinal surface, apart either from the individual characteristics of the plates independently or from other special details, are amply sufficient to distinguish the form from its nearest ally, *Metopaster Parkinsoni*.

5. METOPASTER UNCATUS, *Forbes*, sp. Pl. XI, figs. 3 *a*, 3 *b*; Pl. XIV, figs. 1 *a*—3; Pl. XV, figs. 1 *a*, 1 *b*.

GONIASTER (GONIODISCUS) UNCATUS, *Forbes*, 1848. Memoirs of the Geological Survey of Great Britain, vol. ii, p. 472.

— — — *Forbes*, 1850. In Dixon's Geology and Fossils of the Tertiary and Cretaceous Formations of Sussex, London, 4to, p. 331, pl. xxi, figs. 4, 5, 8.

— — — *Morris*, 1854. Catalogue of British Fossils, 2nd ed., p. 81.

ASTROGONIUM UNCATUM, *Dujardin and Hupé*, 1862. Hist. Nat. Zooph. Échin. (Suites à Buffon), p. 399.

GONIASTER (GONIODISCUS) UNCATUS, *Forbes*, 1878. In Dixon's Geology of Sussex (new edition, Jones), p. 365, pl. xxiv, figs. 4, 5, 8.

Body rather small, or at most only of medium size. General form depressed. Abactinal surface flat, or with a tendency for the rays to be slightly directed upward at the extremities. Actinal surface slightly convex. Marginal contour pentagonal, with the sides slightly lunate, though the curvature is often more or less flattened; the extremity of the rays is only slightly produced. The major radius measures about one-third more than the minor radius, the major dimension being proportionately rather greater in large tests than in small ones. The margin is thick, and the lateral wall has more of a precipitous than a rounded character, although the infero-marginal plates are well rounded on the actinal surface.

The supero-marginal plates are three in number, counting from the median interradial line to the extremity, or six from the tip of one ray to the tip of the adjacent ray, exclusive of the odd terminal or " ocular " plate in each case. They form a broad border to the abactinal area of the disk, of uniform breadth throughout, which measures about 7 mm. at the median interradial line in an example whose major radius measures 36·5 mm. and minor radius 27 mm. (Pl. XIV, fig. 2 a). Excepting the ultimate paired plates, the four intermediate supero-marginal plates on each side of the disk are of equal size, the breadth being about once and a half the length, the actual measurements in the specimen under notice being length 5 mm., and breadth 7 mm. The abactinal surface of these plates is tumidly convex, while the lateral wall is plane and vertical, and the abactinal tumidity commences abruptly at a little distance from the adcentral margin of the plate, which leaves a small level area at the rounded end of the plate abutting on the abactinal plates or paxillar tabulæ. On the surface of this level band near the adcentral margin, and forming a more or less definite series running parallel to it, are three or four irregular tubercular eminences or granules, but very indistinct and more or less weatherworn (see Pl. XIV, fig. 2 c). The entire margin of the plate is surrounded by a very narrow depressed border, with very fine, closely crowded, uniserially disposed punctations, upon which a small miliary granulation was previously borne. The whole general superficies of the plate is smooth and weatherworn in every example I have seen. The height of the plates as seen in the margin is as great as or even slightly greater than their length, and the prominently tumid character of the plates abactinally causes them to appear in the lateral view somewhat like truncate cones abruptly rounded (see Pl. XIV, fig. 2 b).

The ultimate paired plate is larger and longer than any of the other plates, and is of a different and very peculiar shape. It is subtriangular in form as seen from above, produced and pointed at the extremity, and to a certain extent recalls the form of a ploughshare or coulter in consequence of a peculiar nipped-in appearance caused by the extension of a tumid region which runs parallel to the outer margin and rises abruptly from a level area which occupies the inner half of the plate along the margin touching the corresponding plate of the adjacent ray, —the line of junction of the two plates coinciding with the median radial line. The length of the ultimate plate is nearly twice that of the other marginal plates, measuring 9·25 mm. in an example whose major radius is 36·5 mm. and minor radius 27 mm. As seen in the marginal view of the test, the ultimate is not higher or more tumid than the other marginal plates (see Pl. XIV, fig. 2 *b*). The outer margin of the ultimate plate has a slight concave curvature, and the inner margin adjacent to the corresponding plate is curved convexly towards the proximal end of the plate. In consequence of this rounding the two ultimate plates in a pair do not unite throughout their entire length, but are separated by a small notch at the end of the suture adjacent to the abactinal paxillar area of the disk. On the small level area of the ultimate plate are a number of small irregular tubercular eminences; four or five larger than the others form a sort of series parallel to the rounded margin, and a longitudinal series of eight or nine much smaller miliary granules run along the flank of the longitudinal tumidity of the plate; and several additional granules of intermediate size may be present in the space between the two series just described. Excepting these granules, the surface of the ultimate plate is smooth like that of the other supero-marginal plates.

The abactinal area of the disk within the boundary of the marginal plates is covered with small subregular plates or paxillar tabulæ, an hexagonal form predominating. The plates in the median interradial areas are much larger than the other plates on the disk, and a marked diminution in size in all the plates takes place as they approach the margin. All the plates have their surface marked with a very fine granulation. Small entrenched pedicellariæ are occasionally present, but there appear to have been very few.

The primary basal plates are larger than any of the other abactinal plates. They are well seen in an example from the Upper Chalk of Kent, in which the inner side of the abactinal wall is exposed by the removal of the actinal floor and ambulacral plates. This specimen, which is preserved in the British Museum, and bears the registration number " 35,496," is drawn on Pl. XI, fig. 3 *a*. The example in question is further interesting in showing that the plates of the radial regions have stellate bases, whereas the larger plates of the interradial regions are sharply hexagonal, and fit closely to their adjacent plates (see fig. 3 *b*).

The madreporiform body is very small, and is subsagittiform or irregularly lozenge-shaped in outline; in the example under notice it is embedded, all except two straight sides, in one large adcentrally placed basal plate (see Pl. XIV, fig. 2 *d*); the two straight sides are bounded each by one large plate. The surface of the madreporite is sculptured by very fine striations, which though more or less wavy are directed subparallel to the adcentral sides of the body.

The position of the madreporite is nearer the centre of the disk than the margin.

The infero-marginal plates are six in number, counting from the median interradial line to the extremity,—that is to say, there are twelve for the whole side of the disk, as against six in the supero-marginal series. The length of the two innermost plates on each side of the median interradial line is subequal to that of the superior series, but the four succeeding infero-marginal plates are much shorter, and together correspond to the large ultimate supero-marginal plate. As seen in the lateral wall of the disk the height of the infero-marginal plates is very little more than one-half that of the superior series, the actual proportion in the specimen under notice being less than three-fifths. The breadth of the plates adjacent to the median interradial line on the actinal surface is 7·5 mm. in an example whose major radius measures 25 mm. and the minor radius 20 mm. The breadth of the marginal border diminishes so rapidly on each side of the median interradial line towards the extremities of the rays that the inner margin of the series of infero-marginal plates of each side of the disk is practically a sector of a circle. The surface of the plates is marked with small widely spaced punctations, and there is a narrow depressed border round the entire margin of the plate which is very finely punctate; the border is broader at the adcentral end of the plate and the adjacent corners than elsewhere.

The adambulacral plates are small, broader than long, and their surface is marked with about three oblique series, and an outer irregular group of fine punctations upon which the spinelets composing the armature of the plates were articulated. There are about six punctations in the innermost ridge or series adjacent to the ambulacral furrow. A few spinelets are still preserved in an example from the Upper Chalk near Bromley, from which this description is taken, which is in the British Museum collection, and bears the registration number, "E 2613." The spinelets are cylindrical, truncated, and do not appear to taper; and their length is about equal to the length of the plates.

The actinal intermediate plates are rhomboid and hexagonal or polygonal in form, and their surface is entirely covered with small equidistant punctations, upon which small uniform granules were previously attached. Remains of this granulation are occasionally to be found preserved in situ, and may be seen in the example referred to in the preceding paragraph. Small entrenched pedicellariæ are occasionally present on the actinal intermediate plates.

Dimensions.—The example figured on Pl. XIV, fig. 2 *a*, has a major radius of 36·5 mm. and a minor radius of 27 mm. The length of the side is about 40 mm. Another specimen, shown in fig. 1 *a* on the same plate, is rather larger, the length of the side being about 45 mm. Much displacement of the plates has occurred in this fossil, and the radial dimensions can only be calculated approximately. The specimen figured by Forbes, which I have unfortunately not been able to trace, is smaller than either of the above. The drawing represents a test with the major radius measuring about 25 mm., the minor radius 20 mm., and the length of the side about 28 mm. The figure of the actinal aspect of this example is reproduced on Pl. XIV, fig. 3, of this memoir.

Locality and Stratigraphical Position.—This is a well-known Upper-Chalk form. It is recorded by Forbes from Kent, Sussex, and Wiltshire. Specimens in the British Museum bear the localities of Charlton, Gravesend, Bromley, and " Kent."

History.—This species, primarily described by Forbes in his memoir ' On the Asteriadæ found fossil in British Strata,' was first figured in Dixon's ' Geology and Fossils of the Tertiary and Cretaceous Formations of Sussex,' London, 1850. The example which I consider must undoubtedly have been the type (Dixon, pl. xxi, figs. 4 and 5) I have hitherto unfortunately not been able to find. It is stated by Forbes to have formed part of the late Mr. Pearce's collection, and to have been obtained from the Wiltshire Chalk. A badly drawn fragment (Dixon, pl. xxi, fig. 8) is preserved in the British Museum, but I do not consider it to be correctly referred to this species. In my opinion it belongs to a distinct species, *Metopaster cingulatus*, of which a description is given at p. 53.

The examples of this species which have been drawn on Pl. XIV are preserved in the British Museum.

6. METOPASTER SUBLUNATUS, *Forbes*, sp.

GONIASTER (GONIODISCUS) SUBLUNATUS,	*Forbes*, 1848.	Memoirs of the Geological Survey of Great Britain, vol. ii, p. 472.
— — —	*Forbes*, 1850.	In Dixon's Geology and Fossils of the Tertiary and Cretaceous Formations of Sussex, London, 4to, p. 331.
— — —	*Morris*, 1854.	Catalogue of British Fossils, 2nd ed., p. 81.
ASTROGONIUM SUBLUNATUM,	*Dujardin and Hupé*, 1862.	Hist. Nat. Zooph. Échin. (Suites à Buffon), p. 399.
GONIASTER (GONIODISCUS) SUBLUNATUS,	*Forbes*, 1878.	In Dixon's Geology of Sussex (new edition, Jones), p. 365.

This species was described by the late Professor Edward Forbes in Dixon's 'Geology of Sussex' in 1850 in the following words:

"Body pentagonal, with gently lunated sides. Superior intermediate marginal plates four, nearly equal, plain, smooth, or minutely punctate. Inferiors similar. Superior oculars mitrate, large, triangular, acuminated. Ossicula of disc punctate.

"Easily distinguished from the last species [*uncatus*] by its flattened marginals and from the next [*Hunteri*] by its lunated sides.

"Mus. Bowerbank, from the white chalk; also in the collection of the Geological Survey" (op. cit., p. 331).

No figure of the species has ever been published, and no record exists as to the specimen or specimens seen or used by Forbes as type. I have been unable to find any example from the Bowerbank Collection to which this name has been or can be applied; and the only examples which I have seen referred to this species at all are four fragmentary specimens now preserved in the Museum of the Geological Survey in Jermyn Street, and not more than two of these could have been in that collection in 1850.

After a careful study of these specimens I am bound to confess that I find no character by which they can be separated from *Metopaster uncatus*; and if the diagnosis is correct I am led to consider that I have certainly not seen Forbes's type. The *supero*-marginal plates in the specimens in question are *not* flattened, and cannot be said to differ in character from those of *Metopaster uncatus*.

A possible explanation suggests itself in the supposition that Forbes inadvertently mistook the actinal for the abactinal surface of the disk, a mistake which might easily be made by a less careful observer than the author of this species when dealing with a badly preserved fragment. If, however, what is really the actinal surface has been described as the abactinal surface the difficulty is practically solved, for the infero-marginal plates in the fragments under notice *are* "plain, smooth, or minutely punctate." That this is not an improbable explanation I would submit the following facts :—(1) that in the original diagnosis of 1848[1] Forbes states that the infero-marginal plates are unknown; (2) that in the diagnosis of 1850, given above, the infero-marginal plates are stated to be "similar" (to the supero-marginals); and (3) that notwithstanding these statements all the examples in the Geological Survey Collection are essentially actinal presentments of the disk, and therefore the infero-marginal plates are the plates

---

[1] The following is the diagnosis in full :—"G. corpore pentagonali, lateribus lunatis; *ossiculis lateralibus superioribus* 4, subæqualibus, planis, minutissime punctatis; *inferioribus? Ossiculis ocularibus superioribus* magnis, triangularibus, mitratis, tumidis, acuminatis." ('Mem. Geological Survey of Great Britain,' vol. ii, p. 472.)

conspicuously available for description. To my mind it follows with little doubt either that Forbes has described the infero-marginal plates as supero-marginals, or else that I have not seen his type specimens.

With my reverence for all that Forbes has written I prefer to leave the species as described by him, together with the record of these remarks, rather than strike a ruthless pen through any species created by so careful and accurate a worker. Time will pronounce the verdict.

I consider it quite unnecessary to figure any of the fragments, here referred to, as being in the Jermyn Street Collection. The differences they present as compared with a series of *Metopaster uncatus* do not in my opinion amount to even varietal rank, and are confined to the slightly less developed tumidity of the abactinal surface of the supero-marginal plates, and to the external contour of the mitrate ultimate paired supero-marginal plates, which is slightly convex marginally rather than incurved to produce the characteristically claw-shaped form of *Metopaster uncatus*. These are, in my opinion, merely individual differences.

7. METOPASTER CINGULATUS, *Sladen.* Pl. XIV, figs. 4 *a*—4 *d*.

Body of small size. General form depressed. Abactinal surface slightly concave, actinal surface flat. Marginal contour pentagonal, with the sides slightly lunate. The rays are not produced beyond the contour of a true pentagon, and the radial angles are not rounded. The major radius is proportional to the minor radius as 100 : 77·5, the actual dimensions in the example described being, major radius 20 mm., minor radius 15·5 mm , approximately. The margin is thick, and though well rounded has more or less of a precipitous character.

The supero-marginal plates are three in number, counting from the median interradial line to the extremity, or six from the tip of one ray to the tip of the adjacent ray, exclusive of the odd terminal or " ocular " plate in each case. They form a broad border to the abactinal area of the disk of uniform breadth through-out, which measures about 6·5 mm. at the median interradial line in an example whose diameter (R + r) measures from about 36 mm. to 37 mm.

Excepting the ultimate paired plates, the four intermediate supero-marginal plates on each side of the disk are approximately equal in size, the breadth being more than twice and a half the length, the actual dimensions in the example under notice being length 2·25 mm., and breadth 6·5 mm., in the plate adjacent to the median interradial line. The abactinal surface of these plates is distinctly tumid, a subtubercular eminence rising in the median third of the abactinal area of the plate. The slope of the tumidity descends gradually on the outer side, and

8

merges in the rounding of the high lateral wall of the plate. On the level area of the plate, which is consequently the inner or adcentral part of the surface, are several low tubercular eminences of irregular shape and disposition, but which appear to assume a more or less distinct biserial arrangement at right angles to the adcentral margin of the plate. They appear to be enlarged irregular granules, and in all the examples I have examined they have become to a certain extent ill-defined, owing either to growth or to weather-wearing. The general character of the ornamentation is shown in Pl. XIV, fig. 4 c. The entire margin of the plate is surrounded by a very narrow depressed border, which is very minutely punctate, probably only in a single lineal series. The general superficies of the plate beyond the ornamentation mentioned is smooth, as if weatherworn, in all the examples I have seen, but in some specimens there appear to be traces of a more or less granulous character, and in some instances suggest the impression that an ornamentation similar to that noticed in *Mitraster rugatus* was probably present on at least a part of the surface of the plate. The height of the supero-marginal plates as seen in the margin is greater than their length, and their prominent abactinal tumidity has a distinctly conical character from this point of view.

The ultimate paired plate is fully twice as long as the other supero-marginal plates measured on the outer margin, and its breadth is equal to that of the adjacent supero-marginal. It is triangular in form, and the line of junction with the companion ultimate plate of the adjacent side coincides with the median radial line. The actual dimensions in the example under description are, length 6·2 mm., breadth about 6 mm. As seen in the lateral view of the disk the ultimate plates are distinctly tumid abactinally (see Pl. XIV, fig. 4 b). The abactinal surface of the plate is ornamented by a number of miliary tubercles or granules, more or less serially disposed parallel to the margin adjacent to the companion plate, and more numerous at the adcentral end of that margin. Beyond this the surface of the ultimate plate is smooth, like that of the other supero-marginal plates.

The abactinal area of the disk within the boundary of the marginal plates is covered with hexagonal or polygonal plates or paxillar tabulæ, which are small in size generally, excepting the primary apical plates, which are comparatively very large. All the plates have their surface covered with a fine, uniform granulation. The primary apical plates have a small central area of low elevation, not higher than if a number of granules had become coalesced—the structure being suggestive of a tubercle in process of disappearance,—in other words, the scar left by a tubercle which had existed in an earlier stage of growth (see Pl. XIV, fig. 4 d).

Dimensions.—The example figured on Pl. XIV, fig. 4 a, has the following measurements :—major radius, 20 mm.; minor radius, 15·5 mm; length of the side, about 22 mm. The thickness of the margin is 7 mm. in a specimen whose side is 23·5 mm.

Locality and Stratigraphical Position.—The specimen described above and figured on Pl. XIV, fig. 4 *a*, is from the Upper Chalk, near Bromley. It is preserved in the British Museum, and bears the register-number " 46,776." Other examples which I refer to the same species are also in the National Collection, and one, if not more, was obtained from the same locality.

Remarks.—A fragmentary example which was figured in Dixon's ' Geology and Fossils of the Tertiary and Cretaceous Formations of Sussex,' pl. xxi, fig. 8, and was referred by Forbes to *Goniaster* (*Goniodiscus*) *uncatus*, now forms part of the collection in the British Museum, and bears the registered number " E 2577." This specimen appears to me to belong to the present species, and not to the true *Metopaster uncatus*. I can scarcely think that Forbes would have intentionally ranked the two forms as belonging to one species, and I have not thought it needful on the sole grounds of this fragment having been figured to regard *Metopaster cingulatus* as a dismemberment of Forbes' *Metopaster uncatus*.

Metopaster cingulatus is readily distinguished from *Metopaster uncatus* by its very short broad supero-marginal plates, by their more limited and more conical tumidity, by their greater height as seen in the margin, as well as by their different ornamentation. The ultimate plates are triangular, and do not present the peculiar form characteristic of *Metopaster uncatus*. In many respects *Metopaster cingulatus* appears to hold an intermediate position between *Mitraster rugatus* and *Metopaster uncatus*.

8. METOPASTER CORNUTUS, *Sladen*. Pl. XIV, figs. 5 *a*—5 *d*.

Body of small size. General form depressed. Abactinal surface slightly concave, the extremity of the rays being directed slightly upward. Actinal surface slightly convex. Marginal contour pentagonal, with the sides slightly lunate and the extremity of the rays slightly produced. The major radius is proportional to the minor radius as 100 : 74, the actual dimensions in the example described being, major radius about 19 mm., minor radius about 14 mm. approximately. The length of the side is 22·5 mm. The margin is moderately thick and apparently well rounded.

The supero-marginal plates are only two in number, counting from the median interradial line to the extremity, or four from the tip of one ray to the tip of the adjacent ray, exclusive of the odd terminal or " ocular " plate in each case. They form a broad border to the abactinal area of the disk. The supero-marginal plate adjacent to the median interradial line is broader than long, the dimensions in the example described being length about 3·25 mm., and breadth about 5·75 mm. The abactinal surface of this plate is slightly convex in the median line of breadth. The entire margin of the plate is surrounded by a very narrow depressed border with very minute punctations, closely crowded and uniserially disposed near the inner edge of the border, upon which a small miliary granulation was previously borne. The general superficies of the plate is covered with minute punctations irregularly disposed. The height of the two intermediate supero-marginal plates as seen in the lateral view of the margin (Pl. XIV, fig. 5 b) is scarcely half their length, and is less than half the height of the underlying infero-marginal plates.

The ultimate paired plate is very much larger than the small intermediate plate just described, being nearly three times as long, the actual measurement being about 9·25 mm. in the example under notice. It is of a peculiar form, being shaped somewhat like an irregularly formed elytron of an insect as seen from above, the proximal margin being truncate where it joins the companion intermediate plate, and the distal extremity obtusely rounded (Pl. XIV, fig. 5 c). At a point situated three-fourths of the distance from the proximal to the distal extremity rises an abrupt truncate teat-like eminence. The peculiar form of the plate as seen in the lateral view of the margin and the character of the eminence just described will be better appreciated by reference to Pl. XIV, fig. 5 d, than by verbal description. The superficies of the ultimate plate is studded with rather widely spaced punctations disposed in groups, which do not extend over the whole of the area which falls in the lateral wall of the disk. As seen in the marginal view of the test the ultimate is very much higher than the intermediate marginal plates (see Pl. XIV, fig. 5 b).

The odd terminal plate is unknown to me, no trace being preserved in the example under notice.

In like manner the whole of the abactinal plating within the boundary of the supero-marginal plates has been removed, excepting only a few isolated fragments of plates out of position.

The infero-marginal plates are large and high. Four are preserved between the median interradial line and the extremity, but the series is incomplete, probably one or more being wanting at the distal end of the series. This would increase the number eight now preserved on the whole side to ten or perhaps twelve. The two infero-marginal plates on each side of the median interradial line are much larger and higher than the other plates. The plate adjacent to the median interradial line measures about 4 mm. in length and 4 mm. in height; the next plate 5 mm. in length, and from 3·5 mm. to 3·75 mm. in height. The surface of the plates is marked with small widely spaced punctations, and there is a narrow depressed border round the entire margin of the plate, which is very finely punctate.

The actinal area is unknown to me.

History.—The fossil delineated on Pl. XIV, fig. 5 *a*, was drawn by Mr. A. H. Searle under Dr. Wright's instructions, but I regret that I have not been able to find any trace of the specimen. I am therefore led to believe that the type belonged to Dr. Wright's private collection, which has been distributed since his lamented death. Knowing by experience the extreme care and fidelity which characterise all Mr. Searle's work, I have ventured to describe the species from his drawings alone, for it seemed undesirable to leave such an interesting form without notice; and I am hopeful that the publication of the figure and the description of its characteristic features will lead to the detection of the type. I am unable to give any information as to the locality or stratigraphical position from which the fossil was obtained.

Remarks.—The rather small size of this example and the small number of supero-marginal plates—only four for the whole side of the disk—would not unnaturally suggest at first sight that this was possibly an immature form. After careful study, however, I do not consider such to be the case, or at any rate I am unable to regard the fossil under notice as the young of any of the species with which I am acquainted. The large and characteristically developed ultimate supero-marginal plates in conjunction with the presence of a normal number of infero-marginal plates, together with the fact that the size of the test is not less than that of another perfectly characterised species, lead me to rank this as a distinct species with little hesitation. The general proportions as well as the character of the different plates, and the facies of the form as a whole, appear to me to fully warrant this view.

Genus—MITRASTER, *Sladen.*

(Μίτρα = a broad belt, or girdle.)

Body depressed and cycloid, or cyclo-pentagonoid in contour. Marginal plates with co-ordinated granulose elevations and punctations, and a surrounding narrow depressed border with very minute punctations for the articulation of setæ. Supero-marginal plates few in number and all subequal in size, forming a broad uniform border to the disk. Abactinal area covered with polygonal plates, some of which may have stellate or substellate bases, and upon the tabulæ are borne small, more or less co-ordinated granules. Infero-marginal plates more numerous than the supero-marginal series, and decreasing in size as they approach the extremity; the surface marked with punctations, which may be co-ordinated in a similar manner to that of those on the supero-marginal series, and may be associated with granulose elevations. Actinal intermediate plates small, polygonal, covered with uniform, crowded, shallow punctations, upon which granules were originally borne. Armature of the adambulacral plates arranged in longitudinal lines, which may be slightly oblique. Small entrenched pedicellariæ may be present occasionally on the actinal intermediate plates.

Mitraster is characterised by its cycloid contour, by the equality in size of the supero-marginal plates, which do not diminish towards the extremity, and by the character of the ornamentation of the marginal plates, especially of the superior series.

The main characters which distinguish this and the two preceding genera are distinctly relative, and may be here conveniently compared. In *Pentagonaster* the rays are more or less produced, the supero-marginal plates are more or less numerous, and decrease in size as they approach the extremity of the ray, and are devoid of a marginal border of setæ. In *Metopaster* the rays are very slightly produced, the contour being pentagonal and only slightly extended. The supero-marginal plates are few in number, and do not decrease in size as they approach the extremity, the ultimate paired plate being larger than the others. All are furnished with a marginal border of setæ. In *Mitraster* the contour is cycloid almost to the obliteration of the pentagonal form. The supero-marginal plates are few in number, but neither decrease nor diminish in size, being subequal throughout; and they are furnished with a marginal border of setæ.

I consider that these differences indicate structural characters of sufficient morphological significance to render the forms presenting them worthy of recognition as distinct genera.

1. MITRASTER HUNTERI, *Forbes*, sp.　Pl. IX, figs. 3 *a*—3 *e*; Pl. XII, figs. 3 *a*—3 *e*; Pl. XV, figs. 3 *a*—5 *b*.

GONIASTER REGULARIS,	*Mantell*, 1844. Medals of Creation, vol. i, p. 335, lign. 73 (*non* Linck).
GONIASTER (GONIODISCUS) HUNTERI,	*Forbes*, 1848. Memoirs of the Geological Survey of Great Britain, vol. ii, p. 471.
—　　—　　—	*Forbes*, 1850. In Dixon's Geology and Fossils of the Tertiary and Cretaceous Formations of Sussex, London, 4to, p. 331, pl. xxi, fig. 1.
—　　—　　—	*Morris*, 1854. Catalogue of British Fossils, 2nd ed., p. 81.
ASTROGONIUM HUNTERI,	*Dujardin and Hupé*, 1862. Hist. Nat. Zooph. Échin. (Suites à Buffon), p. 399.
GONIASTER (GONIODISCUS) HUNTERI,	*Forbes*, 1878. In Dixon's Geology of Sussex (new edition, Jones), p. 365, pl. xxiv, fig. 1.

Body of small size. General form depressed. Abactinal surface flat and apparently depressed within the boundary of the supero-marginal plates, the gibbosity of the latter and their centrally sloping surface giving a concave character to the area generally in fossil examples, which was probably always more or less present. Actinal surface flat or with a tendency to be slightly convex. Marginal contour subcircularly pentagonoid, the sides being slightly convex or bulging outward. The major radius is proportional to the minor radius as 100 : 90·47 approximately; and the rays are not produced, the angles of the pentagon being obtusely rounded. Margin thick and well rounded, the slope being more gradual on the actinal surface.

The supero-marginal plates are three in number counting from the median interradial line to the extremity, or six from the tip of one ray to the tip of the adjacent ray, exclusive of the odd terminal or " ocular " plate in each case. They form a broad border to the abactinal area of the disk, of uniform breadth throughout, which measures about 6·3 mm. at the median interradial line in an example whose diameter (R + r) measures about 40 mm. Excepting the ultimate plates all the supero-marginal plates are of equal size and uniform, the breadth being a little more than once and a half the length, the actual measurements in the specimen under notice being length 4 mm. and breadth 6·3 mm., in the plate adjacent to the median interradial line. The length of the plate at the end which falls in the margin of the disk is a shade greater than the adcentral or inner end, and the plates are consequently faintly wedge-shaped, but so slightly that the character is scarcely noticed at first sight. The ultimate plates, however, are distinctly wedge-shaped, the length at the outer margin being a little greater than that of the other marginal plates, while the length of the inner end is rather less—often not more than one-half the length of the same end in the other plates. The breadth of the ultimate plates is the same as that of all the supero-marginal plates ; and the corresponding plates of the two adjacent sides touch one another throughout, the line of junction falling in the median line of the ray. The abactinal surface of the plates is distinctly convex, and the character is more conspicuously emphasised by the plate becoming rapidly gibbous on the outer half, the outer side of the eminence forming the rapid bend to the lateral wall of the plate. The height of the plates as seen in the margin is usually equal to, or even rather greater than their length, but may occasionally be less. There is no diminution in height as the plates approach the extremity of the ray, and the ultimate paired plate has a tendency to appear even a trifle higher and more gibbous than the others, but the character is derived probably more from the position in which the plate sometimes is than from an actual increase in size or gibbosity. The abactinal surface of the plates is covered with coarse tuberculiform mammillations which gradually die out before reaching the apex of the gibbosity. In the interspaces between the eminences are small, more or less widely spaced punctations, and these extend over the whole surface of the plate, and are consequently present on the outer portions as well as on the lateral wall. There is a narrow depressed border round the entire margin of the plate, which is very minutely punctate (see Pl. XII, fig. 3 c). In smaller examples there often appears to be only one or two rows of punctations. The ornamentation of the ultimate paired plates is precisely similar to that on the other supero-marginal plates.

The odd terminal or "ocular" plate is very small, and, so far as I can make out, resembles superficially a short truncated cylinder which protrudes somewhat cannon-like from a small triangular space left by the ultimate paired plates similar to what I have already described in *Metopaster*.

The abactinal area of the disk within the boundary of the marginal plates is covered with comparatively large polygonal plates, with closely crowded, rather coarse, uniform granulations, upon which miliary granules or spinelets were previously borne. The primary apical plates are remarkably large, and the plates in the interradial areas are larger than the plates in the radial areas. Of the plates in the radial areas at least the median series and two series on each side have bases of a six-rayed, substellate form. These are admirably seen in an example preserved in the British Museum bearing the register-number "46,772," in which the inner surface of the abactinal floor is exposed.

The madreporiform body, in the examples in which I have detected its presence, appears to be small and subtriangular, and is marked with very coarse striations.

The infero-marginal plates are five in number, counting from the median inter-radial line to the extremity,—that is to say, there are ten for the whole side of the disk as against six in the supero-marginal series. The length of the two innermost plates on each side of the median interradial line is a little greater than that of the corresponding superior plates; the breadth of the plate adjacent to the median interradial line is 7 mm. in the specimen bearing the British Museum register-number "46,766," and is a little greater than the breadth of the corresponding superior plate as seen in the example bearing the British Museum register number "E 2583."

The second plate, counting from the median interradial line, is a little less broad, and the third is slightly more diminished in breadth, and its adcentral margin merges with a sweeping curve into the lateral distal margin, which gives the plate a more or less cuneiform shape. A large portion of the distal lateral margin of this plate abuts on the adambulacral plates. The fourth infero-marginal plate is very small and triangular in form, with the apex directed adcentrally, and with one side abutting entirely on the adambulacral plates. The fifth plate is smaller still. In consequence of the triangular shape of the fourth and fifth plates the greatest length of the third plate is opposite the apex of the fourth plate, and the length of the third plate gradually diminishes up to the outer margin. The breadth of the third plate is 5·6 mm., and that of the fourth plate is only 3·2 mm.

The surface of the infero-marginal plates, which is more or less plain, is ornamented with small, widely spaced, and more or less equidistant punctations, and there is no trace of the tubercular mammillation present on the surface of the supero-marginal plates. On the inner third of the plate the punctations are more closely placed, and they have the appearance of falling into a more or less distinct reticulated arrangement (see Pl. XII, fig. 3 d). A narrow depressed border surrounds the entire margin of each plate, which is very minutely punctate.

The adambulacral plates, which are small, are broader than long, and have their surface traversed by several ridges placed slightly obliquely, but I am unable to define the armature.

The actinal intermediate plates, which are somewhat large in relation to the size of the disk, are rhomboid or polygonal in outline. Their surface is covered with small, closely crowded, uniform pits, upon which miliary granules or spinelets

9

were previously borne (see Pl. XII, fig. 3 *e*). A small entrenched pedicellaria may be present on an occasional plate here and there.

Dimensions.—The example figured on Pl. XII, fig. 3 *a*, has a major radius of about 30 mm., and a minor radius of about 28 mm. The length of the side is about 36 mm. The thickness of the margin is about 10 mm. In the specimen given on Pl. XV, fig. 4 *a*, the major radius measures 21 mm., and the minor radius 19 mm. approximately. The length of the side is about 25 mm.

Locality and Stratigraphical Position.—This species is a characteristic Upper Chalk form. The majority of examples are from Kent and Sussex. A fine series from Bromley is preserved in the British Museum.

History.—This species was referred to by Dr. Mantell under the name of *Goniaster regularis*. The specific name was, however, already preoccupied—at least in literature—for an existing starfish; and although the latter is not now recognisable, no doubt can possibly exist that Mantell's fossil species is certainly a different thing from the starfish to which Linck gave the name of *Pentagonaster regularis*.

Forbes was the first to describe the species under the name of *Goniaster* (*Goniodiscus*) *Hunteri*, and his type-specimen, which is figured in Dixon's 'Geology and Fossils of the Tertiary and Cretaceous Formations of Sussex,' pl. xxi, fig. 1, formed part of John Hunter's Collection, now preserved in the Museum of the Royal College of Surgeons. The examples illustrated in the present work are all preserved in the British Museum.

Variation.—There are at least two well-preserved examples in the collection of the British Museum which I consider to be varieties of this species. They are characterised by the presence of only four supero-marginal plates on each side of the disk, exclusive of the odd terminal plates, as against six plates in typical examples. Beyond this difference in number and the relatively greater length of the supero-marginal plates in proportion to their breadth, I can indicate no character worthy of being noted which would distinguish the examples in question from the typical form of *Mitraster Hunteri*. The specimens under notice measure 35 mm. and 30 mm. in diameter (R + *r*) respectively. Both are from the Upper Chalk, one from near Bromley, the other being only labelled " Kent." Figures of the first-mentioned are given on Pl. IX, figs. 3 *a*—3 *e*.

2. Mitraster rugatus, *Forbes*, sp.   Pl. XVI, figs. 3 *a*—5 *d*.

Goniaster (Goniodiscus) rugatus,	*Forbes*, 1848.	Memoirs of the Geological Survey of Great Britain, vol. ii, p. 471.
— — —	*Forbes*, 1850.	In Dixon's Geology and Fossils of the Tertiary and Cretaceous Formations of Sussex, London, 4to, p. 330, pl. xxi, figs. 2, 2*; pl. xxiii, fig. 15.
— — —	*Morris*, 1854.	Catalogue of British Fossils, 2nd ed., p. 81.
Astrogonium rugatum,	*Dujardin and Hupé*, 1862.	Hist. Nat. Zooph. Échin. (Suites à Buffon), p. 399.
Goniaster (Goniodiscus) rugatus,	*Forbes*, 1878.	In Dixon's Geology of Sussex (new edition, Jones), p. 364, pl. xxiv, figs. 2, 2*; pl. xxvi, fig. 15.

Body of small size. General form depressed. Abactinal surface flat. Actinal surface also flat. Marginal contour pentagonal with almost straight sides. The major radius is proportional to the minor radius as 100 : 75. The actual measurements in an example from the Upper Chalk of Gravesend, preserved in the British Museum, being, major radius, 17·5 mm.; minor radius, 13·75 mm. The rays are not produced beyond the contour of a true pentagon, and the radial angles are not rounded. Margin moderately thick, the lateral wall being almost vertical in consequence of the rapid bend from the abactinal and actinal surfaces; the bend to the actinal surface is, however, rather more gradual.

The supero-marginal plates are three in number, counting from the median interradial line to the extremity, or six from the tip of one ray to the tip of the adjacent ray, exclusive of the odd terminal or "ocular" plate in each case. They form a broad border to the abactinal area of the disk, of uniform breadth throughout, which measures 6 mm. at the median interradial line in an example whose diameter (R + r) measures about 31 mm. Excepting the ultimate plates, the supero-marginal plates are approximately equal in size, the breadth being nearly twice the length, the actual measurements in the above-mentioned specimen of 31 mm. in diameter being, length 3·5 to 3·6 mm., and breadth 6 mm. in the plate adjacent to the median interradial line. The corresponding plate in another example has a breadth of 5 mm. and length 2·75 mm., and in a third, breadth 5·5 mm. and length 2·75 to 2·8 mm. In the plate adjacent to the median interradial line the length at the outer and at the adcentral margin is equal, but in the second plate the length at the adcentral margin is greater than at the outer margin in consequence of a slight extension of the plate to make up for the diminution in size of the ultimate plate, the distal lateral margin of the second plate appearing to be slightly hollowed out for the reception of the ultimate plate, and the corresponding second plates of two adjacent rays consequently touch at their adcentral ends. The ultimate plate is triangular or wedge-shaped in form, and its breadth is less than that of the other supero-marginal plates, and the corresponding plates of the two adjacent sides touch one another thoughout, the line of junction falling in the median line of the ray. The ultimate plate is not unfrequently a little longer at the outer margin than the other supero-marginal plates.

The abactinal surface of the plates is regularly convex along the line of breadth, almost resembling the segment of a cylinder, and no special gibbosity is developed. The abactinal area and the lateral area of the plate form a right angle, and the uniting curve is short and abrupt. The height of the supero-marginal plates, as seen in the margin, is less than their length, the height being 2·25 mm. where the length is 2·75 mm. There is no diminution in height as the plates approach the extremity of the ray, and no special prominence is noticeable in the ultimate plate. The abactinal surface of the plates is ornamented with irregular and frequently elongate eminences or ridges, which are usually disposed transversely or in the direction of the length of the plate. Two or three series may be present, as in Pl. XVI, fig. 4, or the linear ridges may be bent at a right angle, or otherwise curved—a character which is probably the result of several eminences being merged together, by which means a peculiar hieroglyphic-like marking is produced, which fancifully resembles Chinese writing to a certain extent (see Pl. XVI, fig. 5 c). This character is emphasised in weathered specimens. The ornamentation covers the whole abactinal area of the plate, but does not extend upon the lateral wall. Sometimes the form of the eminences is more rounded and tuberculiform, as in *Mitraster Hunteri;* this is distinctly the case in one example, and minute punctations are present in the channels which intervene, but it is to be noted that the tubercles and pits cover the *whole* abactinal area of the plate, and that there is no gibbosity, with its outer flank devoid of tubercles, as in *Mitraster Hunteri.* There is a very narrow depressed border round the entire margin of the plate, from which the regular convexity of the plate rapidly rises. The border is very minutely punctate, and there appears to be usually only a single regular lineal series of punctations. The ornamentation of the ultimate paired plates is similar to that on the other supero-marginal plates, but is more rounded and tubercle-like even in an example in which the broad lineal ornamentation occurs on the other plates.

The odd terminal or so-called "ocular" plate must be very small in order to fit the space left by two adjacent ultimate supero-marginal plates; but I am unable to describe the plate, as I have only found a trace of it in one instance, and have not seen a perfect one preserved in any of the examples of the species I have examined.

The abactinal area of the disk within the boundary of the marginal plates is paved with large polygonal plates, whose surface is marked with a very minute, closely crowded uniform granulation, upon which small miliary granules were previously borne. Traces of these granules are still occasionally present in some examples. The abactinal plates in this species are all large in relation to the size of the disk, and the primary apical plates are much larger than any of the others. The primary apical plates and several of the other plates have a central, small, low, irregular eminence, such as might be formed by the merging together and partial grinding down of a number of granules. I am unable to explain this structure, or to consider it as associated with a pedicellarian apparatus, of which I see no trace. It may possibly mark the remains of a tubercle which had existed in an early stage of the animal's life, but had disappeared and been outgrown at a later stage.

The madreporiform body, which is very small, is circularly subtriangular in shape, and is marked with fine regularly radiating striations. In the example under notice two large plates surround two-thirds of the circumference of the madreporite. In this specimen the madreporiform body measures 1·5 mm. in diameter, and the largest primary plate is 3·2 mm. in diameter.

The infero-marginal plates are five in number, counting from the median interradial line to the extremity,—that is to say, there are ten for the whole side of the disk as against six in the supero-marginal series. The length of the two innermost plates on each side of the median interradial line is greater than that of the corresponding supero-marginal plates, consequently the second infero-marginal plate extends a little way beneath the ultimate supero-marginal plate, and the remaining three infero-marginal plates are all under the ultimate supero-marginal plate. In some examples I am inclined to think that probably only four infero-marginal plates were present, counting from the median interradial line to the extremity, in which case the edge of the second and the remaining two plates were under the ultimate supero-marginal plate.

I have unfortunately not seen any example of this species in which the actinal surface of the disk is exposed, I am therefore unable to give the dimensions of the infero-marginal plates on the actinal surface, or to describe their ornamentation. For the same reason I am prevented from offering any remarks on the adambulacral and actinal intermediate plates.

10

Dimensions.—The example figured on Pl. XVI, fig. 3 *a*, has a major radius of about 17·5 mm., and the minor radius is about 13·75 mm. The length of the side is about 20 mm. The fragmentary type-specimen figured by Forbes, which is drawn on Pl. XVI, fig. 5 *a*, was probably about the same size, judged by computation of the half-side.

Locality and Stratigraphical Position.—One of the type specimens figured by Forbes is stated to have been obtained from the Upper Chalk of Wiltshire, but the second specimen, which is now preserved in the British Museum, bears no record of any locality. Forbes also records the species from Kent and Sussex. Authentic examples from the Upper Chalk from Gravesend and "Kent" are preserved in the British Museum.

History.—This species was described by Forbes in his memoir "On the Asteriadæ found fossil in British Strata" ('Mem. Geol. Surv.,' vol. ii, p. 471, 1848), and figures of two examples were given in Dixon's 'Geology and Fossils of the Tertiary and Cretaceous Formations of Sussex,' London, 1850, pl. xxi, figs. 2, 2*; pl. xxiii, fig. 15. The latter specimen is now preserved in the British Museum (register-number "E 2585"), and is illustrated on Pl. XVI, fig. 5 *a*, of the present work. I have not been able to find any trace of the other example figured by Forbes, which originally formed part of the collection belonging to the late Mr. Channing Pearce. Forbes states that it was found in Wiltshire.

Remarks.—Although at first sight the differences between *Mitraster Hunteri* and *Mitraster rugatus* appear well marked, I am not perfectly satisfied as to the species being altogether independent. When the types alone are examined there appears to be no need for any doubt upon this question. But examples occur which are exceedingly difficult to determine on account of presenting features which seem to break down some character which has been regarded as diagnostic of the other species. In illustration of this difficulty I have drawn on Pl. XVI, fig. 3 *a*, an example which I have ranked under *Mitraster rugatus*, but which presents considerable superficial resemblance to *Mitraster Hunteri* in the character of the ornamentation of the supero-marginal plates. I believe, however, that the proportions of the plates, the absence of any abactinal gibbosity, and the extension of the tuberculation over the whole abactinal area constitute, *inter alia*, a justification for regarding the example as *Mitraster rugatus*.

Turning, on the other hand, to a series of *Mitraster Hunteri*, considerable variation is to be noted in the relative length and breadth of the supero-marginal plates, as well as in the amount of gibbosity developed on the abactinal area of the plate. In such an example as that figured on Pl. XV, fig. 3 *a*, the proportions

PLATE IX.

PYCNASTER ANGUSTATUS, *Forbes*, sp.   (Page 21.)

From the Upper Chalk.

FIG.

1 *a*. Abactinal aspect; natural size.   (Coll. Brit. Mus.)
  b. Lateral view of the margin ; natural size.

METOPASTER PARKINSONI, *Forbes*, sp.   (Page 31.)

From the Upper Chalk.

2 *a*. Actinal aspect ; natural size.   (Coll. Brit. Mus.)
  b. Lateral view of the margin ; natural size.
  c. An infero-marginal plate ; magnified.

MITRASTER HUNTERI, *Forbes*, sp.   (Page 59.)

From the Upper Chalk.

3 *a*. Abactinal aspect of a small example with four supero-marginal plates ;
      natural size.   (Coll. Brit. Mus.)
  b. Actinal aspect of the same ; natural size.
  c. Lateral view of the margin ; natural size.
  d. A supero-marginal plate ; magnified.
  e. An infero-marginal plate ; magnified.

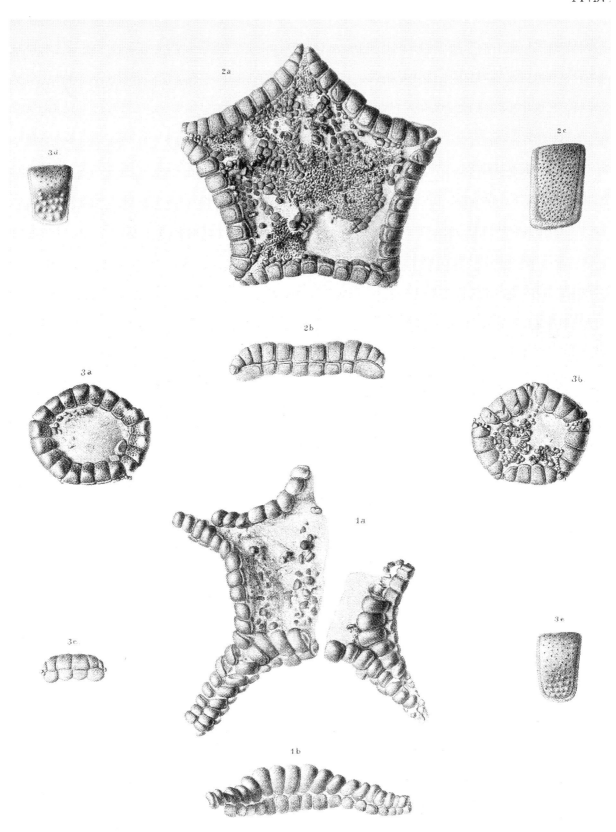

A.H.Searle del.et lith .

Hanhart imp .

PLATE X.

Metopaster Parkinsoni, *Forbes*, sp.   (Page 31.)

From the Upper Chalk.

Fig.

1.   Abactinal aspect; natural size.   (Coll. Brit. Mus.)

2 *a*. Abactinal aspect of another example; natural size.   (Coll. Brit. Mus.)

  b. A supero-marginal plate; magnified 3 diameters.

  c. The madreporiform body; magnified.

  d. An abactinal intermediate plate, with pedicellarian apparatus; magnified.

3 *a*. Actinal aspect of another example; natural size.   (Coll. Brit. Mus.)

  b. Lateral view of the margin; natural size.

  c. Adambulacral plates; magnified.

4 *a*. Actinal aspect of another example; natural size.   (Coll. Brit. Mus.)

  b. An infero-marginal plate; magnified.

  c. An actinal intermediate plate; magnified.

5 *a*. Abactinal aspect of a young (?) example; natural size.

  b. A supero-marginal plate of the same; magnified 4 diameters.

  c. Abactinal intermediate plates; magnified.

Pl. X.

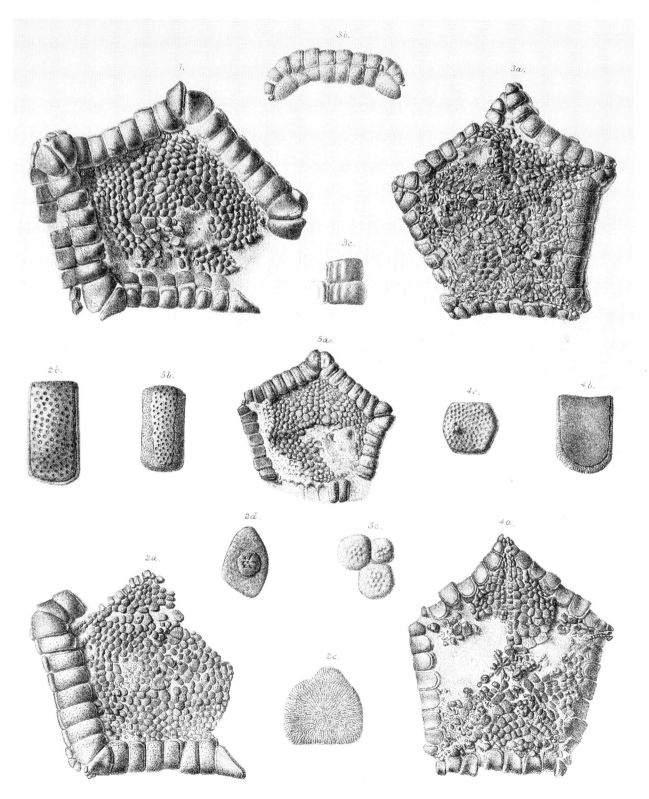

PLATE XI.

Metopaster Parkinsoni, *Forbes*, sp.   (Page 31.)

From the Upper Chalk.

Fig.

1 *a*. Abactinal aspect; natural size.   (Coll. Brit. Mus.)

  b. A supero-marginal plate; magnified 3 diameters.

  c. The madreporiform body; magnified 6 diameters.

2 *a*. Abactinal aspect of another example; natural size.   (Coll. Brit. Mus.)

  b. A supero-marginal plate; magnified.

  c. An abactinal intermediate plate; magnified.

Metopaster uncatus, *Forbes*, sp.   (Page 47.)

From the Upper Chalk.

3 *a*. Actinal aspect of a specimen from which the whole actinal floor has been removed, showing the inner surface of the abactinal floor; natural size. (Coll. Brit. Mus.)

  b. Abactinal intermediate plates, inner surface; magnified.

Pl XI.

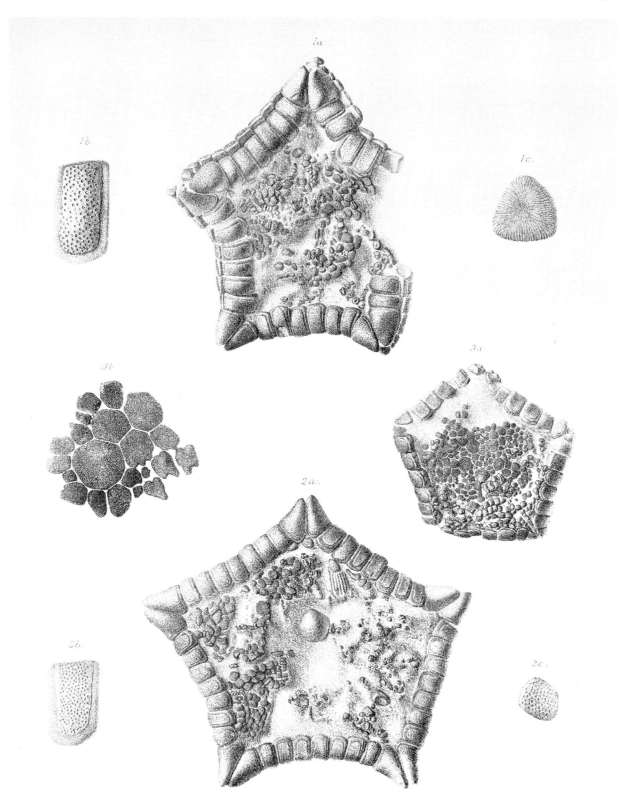

A.H.Searle del.et lith.

Hanhart imp.

PLATE XII.

METOPASTER PARKINSONI, *Forbes*, sp.   (Page 31.)

From the Upper Chalk.

FIG.

1 *a*. Abactinal aspect; natural size.   (Coll. Brit. Mus.)
 b. Lateral view of the margin; natural size.
 c. A supero-marginal plate; magnified.
 d. The madreporiform body; magnified.
 e. Abactinal intermediate plates; magnified.

METOPASTER ZONATUS, *Sladen*.   (Page 45.)

From the Upper Chalk.

2 *a*. Abactinal aspect; natural size.   (Coll. Brit. Mus.)
 b. Lateral view of the margin; natural size.
 c. The madreporiform body; magnified.

MITRASTER HUNTERI, *Forbes*, sp.   (Page 59.)

From the Upper Chalk.

3 *a*. Actinal aspect; natural size.   (Coll. Brit. Mus.)
 b. Lateral view of the margin; natural size.
 c. A supero-marginal plate; magnified.
 d. An infero-marginal plate; magnified.
 e. Actinal intermediate plates; magnified.

CALLIDERMA MOSAICUM, *Forbes*, sp.   (Page 9.)

From the Lower Chalk.

4 *a*. Actinal aspect; natural size.   (Coll. Brit. Mus.)
 b. Lateral view of the margin; natural size.
 c. Mouth-plates; magnified.
 d. An infero-marginal plate; magnified.

Pl. XII.

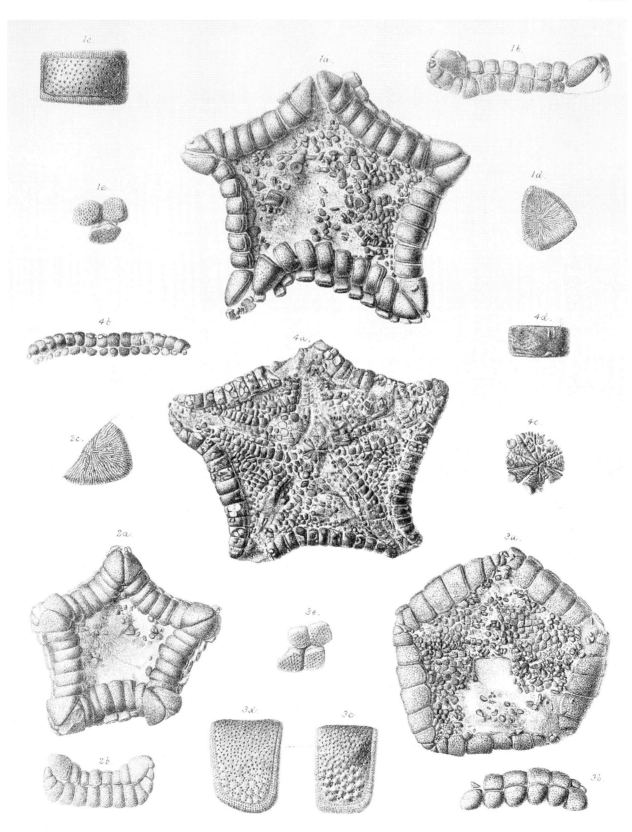

A.H.Searle del.et lith.

Hanhart imp.

PLATE XIII.

PENTAGONASTER MEGALOPLAX, *Sladen*.   (Page 27.)

From the Upper Chalk.

Fig.

1 *a.* Abactinal aspect; natural size.   (Coll. Brit. Mus.)

   b. A supero-marginal plate; magnified 3 diameters.

METOPASTER MANTELLI, *Forbes*, sp.   (Page 38.)

From the Upper Chalk.

2 *a.* Abactinal aspect; natural size.   (Coll. Brit. Mus.)

   b. A supero-marginal plate (much weathered); magnified 3 diameters.

   c. The madreporiform body and surrounding plates; magnified.

   d. Abactinal intermediate plates; magnified 6 diameters.

3 *a.* Abactinal aspect of the example figured by Forbes; natural size.   (Coll. Brit. Mus.)

   b. A supero-marginal plate; magnified 3 diameters.

4 *a.* Actinal aspect of another example figured by Forbes; natural size.   (Coll. Brit. Mus.)

   b. An infero-marginal plate; magnified 3 diameters.

Pl. XIII.

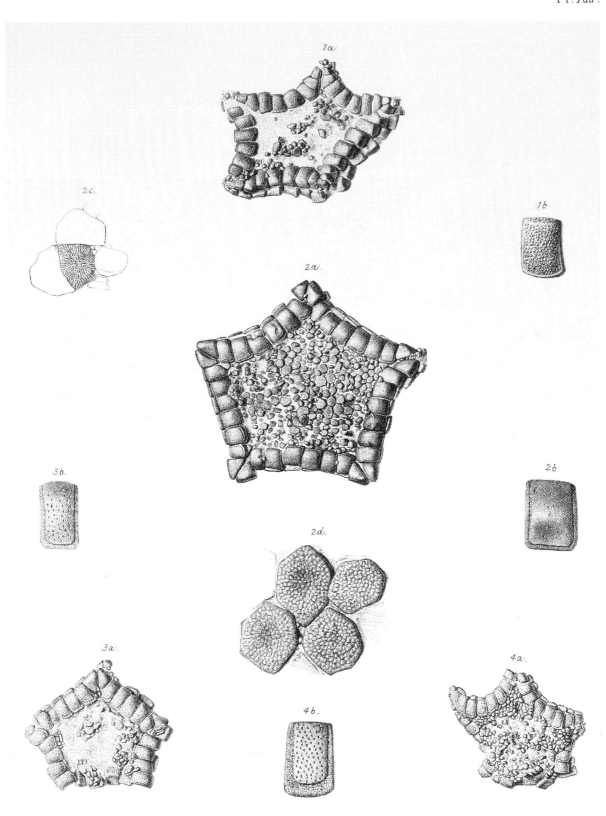

A.H.Searle del.et lith.

Hanhart imp.

PLATE XIV.

Metopaster uncatus, *Forbes*, sp.   (Page 47.)

From the Upper Chalk.

Fig.

1 *a*. Abactinal aspect; natural size.   (Coll. Brit. Mus.).

  b. An ultimate paired supero-marginal plate; magnified.

2 *a*. Abactinal aspect of another example; natural size.   (Coll. Brit. Mus.)

  b. Lateral view of the margin; natural size.

  c. A supero-marginal plate; magnified.

  d. The madreporiform body; magnified 4 diameters.

3. Actinal aspect; copied from the figure given by Forbes in Dixon's 'Geology of Sussex' (pl. xxi, fig. 5).

Metopaster cingulatus, *Sladen*.   (Page 53.)

From the Upper Chalk.

4 *a*. Abactinal aspect; natural size.   (Coll. Brit. Mus.)

  b. Lateral view of the margin; natural size.

  c. A supero-marginal plate; magnified.

  d. An abactinal intermediate plate; magnified.

Metopaster cornutus, *Sladen*.   (Page 55.)

5 *a*. Abactinal aspect; natural size.

  b. Lateral view of the margin; natural size.

  c. An ultimate paired supero-marginal plate seen from above; magnified.

  d. The same seen in profile (lateral view); magnified.

Pl. XIV.

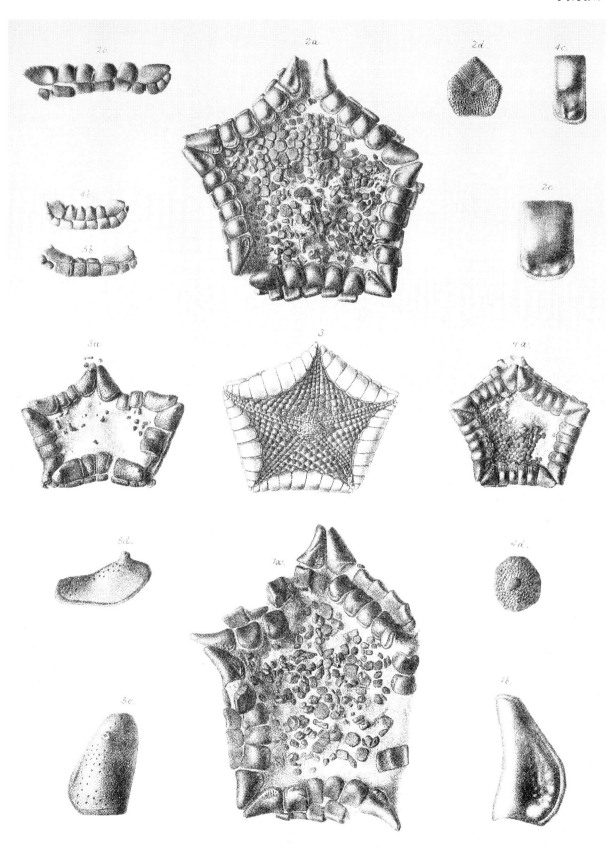

PLATE XV.

METOPASTER UNCATUS, *Forbes*, sp.   (Page 47.)

FIG.

1 *a*. Actinal aspect; natural size.   (Coll. Wright.)

  b. Lateral view of the margin; natural size.

METOPASTER BOWERBANKII, *Forbes*, sp.   (Page 42.)

From the Upper Chalk.

2 *a*. Actinal aspect; natural size.   (Coll. Geol. Survey.)

  b. Lateral view of the margin; natural size.

  c. An infero-marginal plate (weathered); magnified.

  d. Mouth-plates; magnified.

MITRASTER HUNTERI, *Forbes*, sp.   (Page 59.)

From the Upper Chalk.

3 *a*. Abactinal aspect; natural size.   (Coll. Brit. Mus.)

  b. A supero-marginal plate; magnified 3 diameters.

4 *a*. Actinal aspect of another example; natural size.   (Coll. Brit. Mus.)

  b. An infero-marginal plate; magnified 3 diameters.

  c. Extremity of the radial region; magnified 3 diameters.

  d. Abactinal intermediate plates, seen from within; magnified 6 diameters.

5 *a*. Actinal aspect of another example; natural size.   (Coll. Brit. Mus.)

  b. An infero-marginal plate; magnified 3 diameters.

Pl. XV.

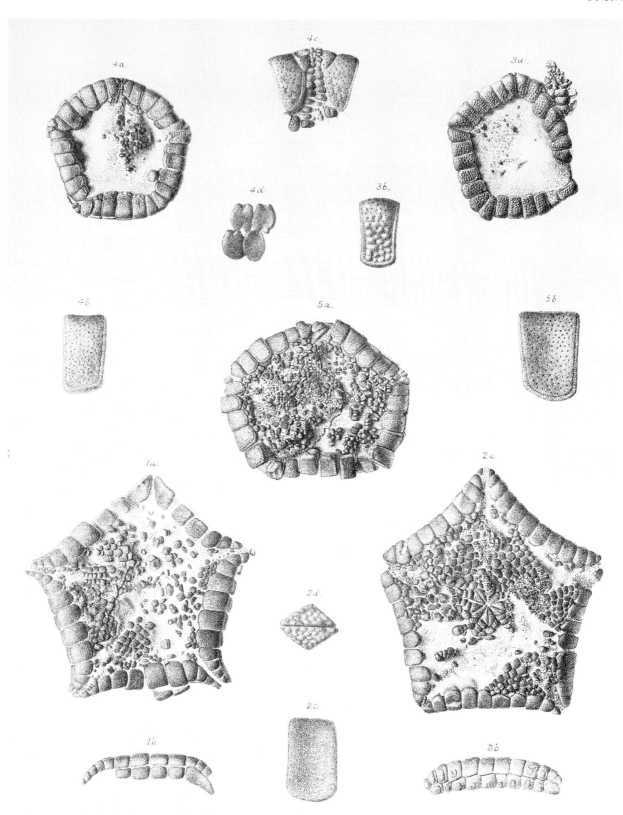

A.H.Searle del.et lith.

Hanhart imp.

PLATE XVI.

METOPASTER BOWERBANKII, *Forbes*, sp.   (Page 42.)

From the Upper (?) Chalk.

FIG.

1 *a*. Abactinal aspect of the example figured by Forbes, natural size.   (Coll. Brit. Mus.)

   b. Lateral view of the margin ; natural size.

   c. The last three supero-marginal plates ; magnified 2 diameters.

   d. Abactinal intermediate plates ; magnified 6 diameters.

METOPASTER PARKINSONI, *Forbes*, sp.   (Page 31.)

From the Upper Chalk.

2 *a*. A pair of ultimate supero-marginal plates and the odd terminal or " ocular " plate ; magnified 2 diameters.   (Coll. Brit. Mus.)

   b. Lateral or marginal view of the same ; magnified 2 diameters.

MITRASTER RUGATUS, *Forbes*, sp.   (Page 63.)

From the Upper Chalk.

3 *a*. Abactinal aspect ; natural size.   (Coll. Brit. Mus.)

   b. A supero-marginal plate ; magnified 3 diameters.

4.   Supero-marginal plates of another example ; magnified 2 diameters.   (Coll. Brit. Mus.)

5 *a*. Abactinal aspect of the example figured by Forbes ; natural size.   (Coll. Brit. Mus.)

   b. Lateral view of the margin ; natural size.

   c. A supero-marginal plate ; magnified 3 diameters.

   d. Profile or sectional view of the margin, showing a supero-marginal and an infero-marginal plate ; magnified 3 diameters.

Pl. XVI.

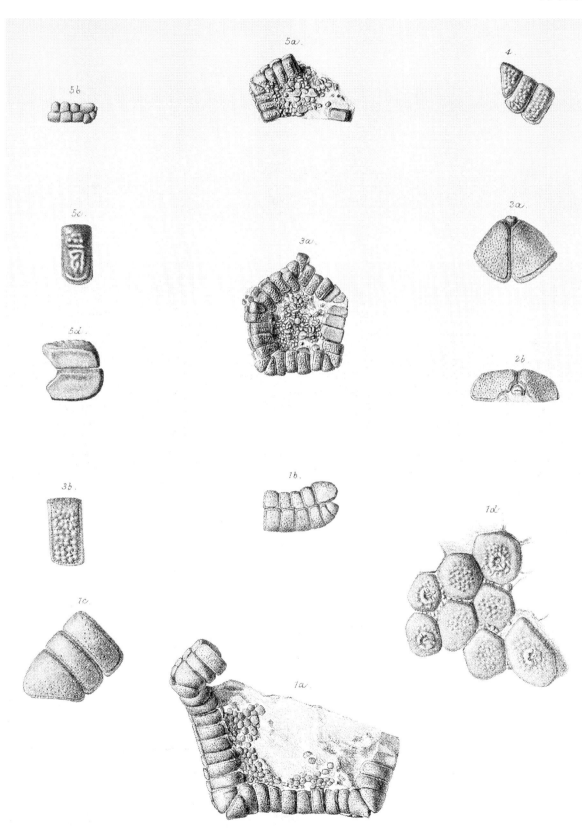

A.H.Searle del.et lith.

Hanhart imp.

Palæontographical Society, 1905.

A MONOGRAPH

ON THE

BRITISH FOSSIL

ECHINODERMATA

FROM

THE CRETACEOUS FORMATIONS.

VOLUME SECOND.
THE ASTEROIDEA.

BY

W. K. SPENCER, B.A., F.G.S.

PART THIRD.
PAGES 67—90; PLATES XVII—XXVI.

LONDON:

PRINTED FOR THE PALÆONTOGRAPHICAL SOCIETY.

1905.

PRINTED BY ADLARD AND SON, LONDON AND DORKING.

of length to breadth are in the case of the interradial supero-marginal plates as 5·6 mm. is to 3·7 mm. Further, the tuberculation in this specimen may or may not extend over the whole of the abactinal area, the variations being on adjacent plates, and the abactinal gibbosity is not strongly developed. In the example figured on Pl. IX, fig. 3, the proportions of length to breadth in the case of the interradial supero-marginal plates are as 4·5 mm. is to 3·6 mm.; the granulated areas more generally stop short of the distal edge of the supero-marginal plate and the abactinal gibbosity is well pronounced. In view of these considerable variations it is difficult to refer these forms to more than one species.

3. MITRASTER COMPACTUS, *Forbes*, sp. Pl. XVII, fig. 2; Pl. XXVI, figs. 3, 3 *a*, 3 *b*, 3 *c*.

GONIASTER COMPACTUS, *Forbes*, 1848. Mem. Geol. Surv. Gt. Brit., vol. ii, p. 468.
— — *Forbes*, 1850. In Dixon's Geology and Fossils of the Tertiary and Cretaceous Formations of Sussex, London, p. 333, pl. xxii, fig. 3.
ASTROGONIUM COMPACTUM, *Dujardin et Hupé*, 1862. Hist. Nat. Zooph. Échin. (Suites à Buffon), p. 399.
GONIASTER COMPACTUS, *Forbes*, 1878. In Dixon's Geology of Sussex (new edition, Jones), p. 366, pl. xxii, fig. 3.

Specific Characters.—Outline pentagonal, slightly cycloidal. Eight supero-marginal plates on each side of the pentagon. Supero-marginal plates form a broad margin, and the breadth of each is about four times its length. Base of ultimate paired supero-marginal plate twice as long as the other, more proximal, supero-marginal plates. Ten corresponding infero-marginalia.

Material.—Only one specimen of this species is known. This formed a portion of Mr. Willett's collection and is now preserved in the Brighton Museum. It apparently escaped the observation of the late Dr. Wright, for the figure on Plate XVII is copied from that in Dixon's 'Geology of Sussex.' As this figure is slightly inaccurate I have had it redrawn and further details added on Plate XXVI.

Description.—The dorsal surface of the disc is covered with a number of small, subequal, closely-fitting plates. It is considerably sunk in the specimen known.

The supero-marginalia bounding the disc form a uniform margin 5·15 mm. broad. They are eight in number along each side, exclusive of the odd terminal or ocular plates. The six middle plates are about 1·2 mm. long. Their breadth is rather more than four times their length, a feature which distinguishes them

11

from *Mitraster Hunteri* and *M. rugatus*. A further distinctive feature is the size of the distal paired plate. This plate is triangular. The base of the triangle measures 2·7 mm., giving the plate twice the length of the more proximal plates. The plate is gibbous at its outer extremity as in *M. Hunteri*. All the plates are ornamented with a single or double marginal row of small spinelets. The ocular is a small conical plate barely visible in abactinal view. It fits into notches on the lower surface of the distal paired plates, and is, as usual, notched on its inferior surface for the purpose of protecting the unpaired terminal tube foot.

The median infero-marginal plates are rather longer than the corresponding members of the superior series. The first two, reckoning from the median inter-radial line, are 1·85 mm. long, and 8·5 mm. broad. The third is only 1·8 mm. long and not quite as broad. The fourth has approximately the same length but is subtriangular in form. The fifth is a small triangular plate. Two infero-marginals and a portion of a third are situated underneath the distal paired supero-marginal plate.

The ventro-lateralia visible are small hexagonal plates covered with a fine uniform granulation. The adambulacralia are small oblong plates. The margin of the disc is very abrupt, but the transition from infero-marginalia to the actinal surface is more gradual than that of the supero-marginalia to the upper surface. A number of small granules are irregularly distributed between the plates.

Remarks.—Unfortunately, the specimen is slightly distorted, so that the pronouncedly cycloidal appearance in the figure is partially due to the unnatural position of the marginal plates, which has brought the inferior series into the dorsal view. The supero-marginal plates appear to have been straight and the inferior series but slightly cycloidal. This, together with the large comparative size of the ultimate paired plate, would bring the species very near to the genus *Metopaster*. Forbes remarked upon the fact that it appeared to be intermediate between *Goniaster* (*Metopaster*, Sladen) *uncatus* and *Goniaster* (*Mitraster*, Sladen) *rugatus*. I have therefore considerable doubt as to the validity of the separation of these two genera.

Locality and Stratigraphical Position.—Upper Chalk of Haughton, Sussex.

Genus—COMPTONIA, *Gray.*

Gray, 1840. Annals and Magazine of Natural History, vol. vi, p. 278.
— 1866. Synopsis of Starfishes in the British Museum.

Body depressed, with produced tapering rays. Disc covered abactinally and actinally with numerous polygonal plates which possess a uniform granulation. Marginal plates numerous. Supero-marginal plates equal in number to the infero-marginals, and forming a moderately broad border to the disc. Infero-marginal plates (as well as all other plates) devoid of spines. Radialia present throughout length of ray.

This genus apparently differs from *Stellaster* only in the absence of spines on the infero-marginalia. It is thus similar to, as well as prior to *Ogmaster* (*von Martens*, 1865) and *Dorigona* (*Gray*, 1866).

1. COMPTONIA COMPTONI, *Forbes*, sp. Pl. XVII, figs. 3, 3 *a*, and 3 *b*; Pl. XVIII, figs. 2, 2 *a*, 2 *b*, 2 *c*, 2 *d*.

STELLASTER COMPTONI, *Forbes*, 1848. Mem. Geol. Surv. Gt. Brit., vol. ii, p. 476.
— — *Forbes*, 1850. In Dixon's Geology and Fossils of the Tertiary and Cretaceous Formations of Sussex, pl. xxii, fig. 8, p. 335.
— — *Dujardin et Hupé*, 1862. Hist. Nat. Zooph. Échin. (Suites à Buffon), p. 408.
— — *Forbes*, 1878. In Dixon's Geology of Sussex (new edition, Jones), pl. xxii, fig. 8, p. 368, 370.

Specific Characters.—Disc large and interbrachial arcs wide, giving the disc a distinctly pentagonal appearance. Major radius rather more than twice the length of the minor radius. Arms elliptical in cross section. Large valvate pedicellariæ present.

Material.—Two specimens (the two cotypes) of this species are in existence. One (formerly in the Bowerbank Collection) displays the actinal aspect (Pl. XVII, fig. 3), and is preserved in the British Museum of Natural History (34311). The other (Pl. XVIII, fig. 2), which shows the dorsal aspect, is preserved in the Northampton Museum. This is the specimen figured in Dixon.

Description.—The large pentagonal disc is covered dorsally with numerous small closely-fitting plates. In the radial areas these plates are polygonal and are about 1·8 mm. in diameter. In the interradial areas the plates measure only 1 mm. in diameter and are rhomboidal. All the plates are covered with a fine uniform granulation (Pl. XVIII, fig. 2 a). Upon very many of the plates are valvate pedicellariæ. Post-mortem changes have produced a sinking in of the plates over the interradial areas. Depressions, doubtless due to similar causes, appear in recent forms when dried, as also in *C. elegans*. I have been unable to distinguish either the madreporite or the anus.

The arms are not so much produced as in *C. elegans*.

R : r :: 62 mm. : 29 mm. in the specimen at Northampton.

R : r :: 55·6 mm. : 25·8 mm. in the British Museum (Natural History) specimen.

The width of the arms at the sixth supero-marginal (reckoning from the inter-radius) is 11·5 mm.

The supero-marginalia are oblong in shape. In the interradial areas they are of fairly constant size, measuring 5·2 mm. in breadth and 2 mm. in length. They diminish in size distalwards. They are eighteen in number, and often bear one or more valvate pedicellariæ. The margin is rounded and is about 8 mm. high.

The infero-marginalia are equal in number and similar in appearance to the superior series.

The actinal interradial areas are large and filled proximally with a number of small rhomboidal plates about 1·2 mm. in average breadth. The more distal plates are crowded, smaller, and polygonal in appearance. Traces of a fine granulation are visible.

The adambulacrals are a series of small oblong prominent plates. The largest are about 1·6 mm. in length and 1·2 mm. in breadth. Remains of their armature are still present. The mouth-angle plates are small and but slightly prominent. They also bear traces of armature. Valvate pedicellariæ are scattered apparently irregularly over all these various plates.

Locality and Stratigraphical Position.—Upper Greensand of Blackdown.

Remarks.—Forbes considered this species was equivalent to *Asterias Schultzii*, Roemer.[1] In this latter species, however, the superomarginalia meet across the dorsal surface of the ray, which would disprove Forbes' statement.

[1] Roemer, ' Versteinerungen des Norddeutschen Kreidegebirges,' pl. vi, fig. 21.

2. COMPTONIA ELEGANS, *Gray.* Pl. XVII, figs 4 and 4 *a*.

COMPTONIA ELEGANS,	*Gray*, 1840. Ann. & Mag. Nat. Hist., vol. vi, p. 278.
STELLASTER ELEGANS,	*Forbes*, 1848. Mem. Geol. Surv. Gt. Brit., vol. ii, p. 476.
— —	*Forbes*, 1850. In Dixon's Geology and Fossils of the Tertiary and Cretaceous Formations of Sussex, London, p. 336, pl. xxii, fig. 9.
COMPTONIA ELEGANS,	*Morris*, 1854. Catalogue of British Fossils, 2nd ed., p. 50.
— —	*Dujardin et Hupé*, 1862. Hist. Nat. Zooph. Échin. (Suites à Buffon), p. 408.
STELLASTER ELEGANS,	*Forbes*, 1878. In Dixon's Geology of Sussex (new edition, Jones), pp. 369, 370, pl. xxii, fig, 9.

Specific Characters.—Disc strongly convex, covered with small polygonal plates. Actinal interradial areas large. Arms well produced, the major radius being at least three times as long as the minor radius. Interbrachial arcs paraboloid.

Material.—The specimen figured by Dixon, at that time in the Bowerbank Collection, is now preserved in the British Museum of Natural History (E. 2567). Both dorsal and ventral aspects are exposed. Another specimen showing an impression of the ventral surface exists in the Oxford University Museum.

Dixon's specimen, however, can hardly be the type, since Gray (1840) stated that the specimens described by him were in the British Museum or in the collection of the Zoological Society. Forbes (1848) refers only to specimens in the British Museum and the collection of the Marquess of Northampton. No part of the Bowerbank Collection is known to have come to the British Museum before 1865. The type specimen therefore must be either lost or still unrecognised in the national collection. Since it was never figured it could never be identified with certainty. It is therefore advisable to take the specimen E. 2567 as type.

Description.—The disc is high in the central and radial regions. In the interradial areas, however, post-mortem changes have caused a collapse of the test and the consequent production of deep triangular depressions. The plates covering the disc are minute, polygonal, and closely fitting. The centrale is the only plate of the dorsal surface which is larger or more conspicuous than the remainder; all are covered with a minute uniform granulation of a quite characteristic appearance. The anus is almost central in position. It is surrounded by a circlet of plates, amongst which is the centrale. The madreporite is, as usual, situated in the next (clockwise) interradius, almost halfway between the centrale and the margin. It is a triangular plate, the apex of the triangle being a markedly acute angle.

The arms are well produced. $R : r : : 30$ mm. $+ : 9$ mm. Their breadth at the base is 9 mm. Radialia, adradialia, and dorso-lateralia extend into the base of the arms. The dorso-lateralia soon disappear, but the adradialia persist as far as the seventh or eighth supero-marginal plate. When the adradials disappear the radialia become larger. They are at this point 1 mm. broad and 1·1 mm. long and therefore appear almost square.

The supero-marginalia are oblong plates of curiously uniform size in the portions of the specimen preserved. They are 1·6 mm. long and from 1·2 mm. to 1·3 mm. broad. The infero-marginalia are of the same length and are opposite to the supero-marginalia. In lateral view the supero-marginalia appear higher than the inferior series. Both series are ornamented with a number of small, fine granules which are uniformly distributed over their surfaces.

The ventral surface is concave. The ventro-lateral plates are rhomboidal in the region of the mouth. They become polygonal and crowded as they approach the margin. Some of these plates extend into the base of the arms. Around the edges of the plates spinelets are visible. The spines of the adambulacral plates are still present. Unfortunately, it is not possible to make out their exact distribution. The mouth-angle plates are not prominent.

There is no trace in this species of such valvate pedicellariæ as characterise *C. Comptoni*.

Remarks.—Gray compared this species with *Cœlaster*, Agassiz.[1] The rather vague diagnosis of *Cœlaster* given by Agassiz renders exact identification impossible.

Locality and Stratigraphical Position.—Upper Greensand of Blackdown. Also in the Upper Greensand at Folkestone (observed by Forbes).

[1] Agassiz, 'Annales des Sciences Naturelles,' 1837. Translated in 'Annals and Magazine of Natural History,' vol. 1, 1838.

Genus—NYMPHASTER, *Sladen*, 1885. (See p. 14.)

4. NYMPHASTER RADIATUS, n. sp. Pl. XXV, figs. 1, 1 *a*, 1 *b*.

Specific Characters.—Arms very much produced. R : r : : 150 mm. : 10 mm. Supero-marginalia in contact almost the whole length of arm.

Material.—The only specimen of this species, formerly in the collection of Mr. J. Starkie Gardner, is preserved in the British Museum of Natural History (E. 375). The plates have, unfortunately, disappeared from the disc. Practically all that remains is the greater portion of one arm.

Description.—At the base of the arm the supero-marginalia are oblong. Each measures 2·8 mm. in breadth, 2 mm. in length, and 3 mm. in height. Distally these plates become almost square. They are ornamented by small granules which tend to run together transversally to the length of the arm (Pl. XXV, fig. 1), and articulations for spines and deep depressions for pedicellariæ are also present. They are about twenty-five in number. The infero-marginalia equal in number and size and oppose the supero-marginalia. Further they are about the same height in marginal view.

The breadth of the arm at the fourth supero-marginal is 6 mm.

Stratigraphical Position.—Lower Chalk. Locality uncertain.

Genus—PENTAGONASTER, *Linck.* (See p. 24.)

3. PENTAGONASTER ROBUSTUS, n. sp. Pl. XXI, figs. 2, 2 *a*.

Specific Characters.—Disc covered with small rounded plates. Margin high. Rays short, high and robust. R : r : : 21·5 mm. : 9·9 mm. The supero-marginal plates meet along the median line throughout almost the whole length of the arm, and form a broad border to the disc. Interbrachial arcs paraboloid.

Material.—The only specimen of this species is the one here described, formerly in the Mantell collection and now preserved in the British Museum of Natural History (48085). The locality from which it was derived is stated rather vaguely as Upper Chalk, Sussex. The specimen is somewhat imperfect, consisting only of the dorsal view of three arms and a portion of the disc.

Description.—The disc appears to have been covered on its dorsal surface by a large number of small, rounded, closely-fitting plates. Mostly they are subequal in size and have an approximate diameter of 2 mm. An uncertain number of even smaller granule-like plates exist scattered between these.

Both radialia and adradialia extend into the base of the arms, but only as far as the third supero-marginal plate, counting from the median interradial line. The arms themselves are short and high. The height of the specimen in the interradius is 9·2 mm. From this point the height gradually diminishes to the extremity of the ray, where it is 5 mm.

The supero-marginalia are about twelve in number. They form a broad margin to the disc and rays. Each supero-marginal plate is high, and is very convex dorsally. Hence every plate is very distinct. The six proximal supero-marginals diminish only slightly in size distalward along the ray. The next six, however, diminish much more rapidly. The supero-marginal nearest the inter-radius has the following measurements: height, 5·75 mm.; breadth, 4·5 mm.; length, 3 mm.

The ocular plate has broken away, and there is no trace of a madreporite.

The infero-marginalia alternate with the supero-marginal series. They are not so high and much squarer in appearance. They decrease in size much more rapidly than the upper series (see Pl. XXI, fig. 2 *a*). The infero-marginal plates, nearest the interradius, measure 4·5 mm. high and 3·2 mm. in length. Any orna-ment that may have existed has disappeared from all parts.

Locality and Stratigraphical Position.—Upper Chalk, Sussex.

4. PENTAGONASTER OBTUSUS, *Forbes*, sp.   Pl. XXII, figs. 1, 1 *a*, 1 *b*, 2, 2 *a*, 3 *a*, 3 *b*, 3 *c*, 3 *d*, 3 *e*, 3 *f*, 3 *g*.

OREASTER OBTUSUS, *Forbes*, 1848.   Mem. Geol. Surv. Gt. Brit., vol. ii, p. 468.
—   —   *Forbes*, 1850.   In Dixon's Geology and Fossils of the Tertiary and Cretaceous Formations of Sussex, London, p. 330, pl. xxi, fig. 12.
—   —   *Dujardin et Hupé*, 1862.   Hist. Nat. Zooph. Échin. (Suites à Buffon), p. 389.
—   —   *Forbes*, 1878.   In Dixon's Geology of Sussex (new edition, Jones), pl. xxi, fig. 12, pp. 364, 370.

Specific Characters.—Disc slightly convex. Majority of the plates covering the disc of subequal size and closely set. R : r : : 25 mm. : 12 mm. Extremities of arms obtuse. Interbrachial arcs only slightly rounded, giving the disc a distinctly pentagonal appearance.

Material.—The two extremities of the arms from which Forbes originally described the species when in the Dixon collection, are now preserved in the British Museum (Natural History). They are not on the same slab of chalk as represented by Forbes, but are and probably always have been two independent specimens, E. 5038 (Pl. XXII, figs. 3 *b, c, d*), and E. 5039 (fig. 3 *a*). More complete specimens have since been added to the collection (40400, Pl. XXII, fig. 1 *ex* J. Simmons' Coll., and 35481, Pl. XXII, fig. 2, *ex* H. W. Taylor's Coll.). Two extremities of arms are also preserved in the Brighton Museum.

Description.—The disc is high and distinctly pentagonal. At the edge of the disc the dorsal covering plates are crowded and polygonal in appearance. Towards the centre they become slightly smaller and rounded. The average diameter of these plates is 1·7 mm.

The madreporite is subcentral in position. It is about the same size as the other plates of the disc and is pentagonal in shape (Pl. XXII, fig. 1 *a*).

The arms are stoutly built. A triple row of polygonal plates enters their bases. The adradial series soon disappears, leaving the single radial series, which appears to persist until it reaches that part of the ray which is obtuse. From this point the supero-marginal plates may or may not be adjunct up to the end of the ray. Considerable variation appears to exist as to this point in the single specimen examined. The arms are distinctly obtuse in their distal half. This has given the species its name.

The supero-marginalia form a rather broad border to the disc. There are nine supero-marginals from the median interradius to the extremity of the ray. Towards the end of the ray the plates of this series become narrower, more oblong in appearance, and distinctly convex.

The infero-marginalia are equal in number and situated generally alternating with the supero-marginal series. They are distinctly square in shape, especially at the obtuse extremities of the ray. Their ornament is in some specimens not so coarse as that of the supero-marginals.

The actinal interradial areas are very distinct and are occupied by four series of ventro-lateral plates. As usual, the actinal plates near the mouth are larger and more rhomboidal than the distal plates. Ventro-lateral plates only extend throughout about a quarter of the length of the arm.

12

The mouth-angle plates are not prominent. The adambulacral plates possess a triple row of spines.

Dimensions.—Specimens 35,481 and 40,400.—The greatest width of the ray varies from 8·3 to 6·5 mm., and the least width from 7·2 to 5·7 mm. The supero-marginal plates are 3·2 mm. broad near the interradii.

Specimen E. 5038.—Greatest width of ray 10·5 mm.

Specimen in Brighton Museum.—Greatest width of ray 9·2 mm.

Locality and Stratigraphical Position.—Upper Chalk, Lancing, Sussex, and also from the Upper Chalk of Kent.

Family—PENTACEROTIDÆ (*Gray*) *emend. Perrier*, 1884.

Phanerozonate Asteroids with unequally developed marginal plates, the superior series being frequently masked or hidden in membrane. Abactinal skeleton reticulate. Plates with large isolated tubercles, or spinelets, or granulose, or covered with membrane. Actinal interradial areas with large pavement-like plates which bear unequal-sized granules.

Genus—PENTACEROS, *Schulze*, 1760.

PENTACEROS,	*Schulze*, 1760.	Betrachtung der versteinerten Seesterne und ihrer Theile, Warschau u. Dresden, p. 50.
GONIASTER (pars),	*Agassiz*, 1835.	Mem. Soc. Sci. Nat. Neuchatel, t. i, p. 191.
PENTACEROS,	*Gray*, 1840.	Ann. and Mag. Nat. Hist., vol. vi, p. 276.
OREASTER,	*Müller* and *Troschel*, 1842.	System der Asteriden, p. 44.

Generic Characters.—Form stellate, marginal plates conspicuous, defining the ambitus. Abactinal plates regular, with more or less definite intermediate papular areas. Prominent localised mammillated tubercles or spines present.

All the fossil species of this genus possess intermarginalia, but do not otherwise approach Sladens' genus *Pentaceropsis* which possesses this character. In view of the fact that intermarginalia may occur as a variation in unmistakable recent species of *Pentaceros* this character cannot invalidate the admission to the present genus of the species about to be described.

1. PENTACEROS BULBIFERUS, *Forbes*, sp.   Pl. XX, figs. 1, 1 *a*, 1 *b*, and 1 *c*; figs. 2, 2 *a*,
and 2 *b*; Pl. XXI, figs. 1, 1 *a*, 1 *b*, 3, 3 *a*,
4, 4 *a*; Pl. XXIII, figs. 2, 2 *a*.

OREASTER BULBIFERUS, *Forbes*, 1848.   Mem. Geol. Surv. Gt. Brit., vol. ii, p. 468.
—   —   *Forbes*, 1850.   In Dixon's Geology and Fossils of the
Tertiary and Cretaceous Formations of Sussex,
pp. 328, 329, pl. xxiv, fig. 7.
—   —   *Dujardin et Hupé*, 1862.   Hist. Nat. Zooph. Échin. (Suites
à Buffon), p. 389.
—   —   *Forbes*, 1878.   In Dixon's Geology of Sussex (new
edition, Jones), pp. 363, 370, pl. xxiv, fig. 7.
—   —-   *P. H. Carpenter*, 1882.   Geol. Mag., p. 12.

Specific Characters.—Disc and arms very convex.   The centrale and primary
interradialia large and tuberculiform.   The major radius approximately twice the
minor radius.   Radialia of the arm conspicuous.   Extremities of the arms swollen.

Material.—The specimens figured and described are all preserved in the
British Museum (Natural History).   E. 5040 (Pl. XXI, fig. 1), 40175 (Pl. XX,
fig. 1), 48748 (Pl. XX, fig. 2), and E. 5041 (Pl. XXI, fig. 3), which were bought
from J. Simmons, and 40399 (Pl. XXI, fig. 4), from the collection of E. Charlesworth,
are all labelled as coming from the Upper Chalk of Bromley, Kent, which,
however, seems to be an inexact dealer's locality, probably intentionally mis-
leading.   E. 5042 (Pl. XXIII, fig. 2), also bought of J. Simmons, is labelled
"Upper Chalk, Charlton, Kent."

Other specimens are known in the Sedgwick Museum, Cambridge, Northampton Museum, and Brighton Museum.   Specimens have also been described by
Valette from the South of France.

This seems to be much the commonest as well as the most graceful of the
Chalk Pentacerotidæ.

Description.—The general aspect of the plates of the disc gives this species a
very characteristic appearance, for the five primary interradialia and the centrale
are very prominent.   They have a lobed widely-spreading base, and are swollen
on the upper surface into an almost spherical form.   Their weathered surfaces
are pitted in a very regular manner, the pits indicating the former presence of

granules. Sometimes the granules are still present *in situ*. The pits are separated in the example figured Pl. XXI, fig. 1, on an average rather less than their own diameter apart. There may or may not be a slight margin to the plate. The centrale in a specimen R : r : : 40 : 20 measures 8·5 mm. in diameter. The primary interradialia are rather smaller, being 6·7 mm. in diameter. Radially the most conspicuous plates of the disc are the proximal radialia. They have a very characteristic appearance, their general shape reminding one of a breastplate. The remainder of the plates of the disc are of very various sizes and distributed in a fairly regular manner. The general arrangement of these plates is given in the general account at the conclusion of these volumes.

The madreporite is a conspicuous plate lying at the distal end of a primary interradial. The two neighbouring adradialia are notched for its reception.

The arms are moderately produced, the major radius being about twice the minor radius. Measurements of five specimens give the following :

R	r
40 mm.	20 mm.
35 mm.	17 mm.
50 mm.	25 mm.
50 mm.	25 mm.
50 mm.	20 mm.

At the base of each arm there are five series of plates visible on the dorsal surface—the radialia, adradialia, and supramarginalia. All the plates at the base of the arm overlap. They are of a type which may be derived from the breastplate shape mentioned above. They gradually become narrowed in length and increased in breadth until they are shaped somewhat like an inverted T (Pl. XX, fig. 2 *b*). The granulation is generally confined to the central region of each plate.

The arm about halfway along its length becomes swollen and the plates no longer overlap but are contiguous. They lose their ⊥-shaped form, become almost oblong, and at the same time rather tumid. This is especially noticeable in the case of the radialia. The form of the plates is, however, rarely absolutely regular, but one which is generally derivable from the breastplate shape.

If we examine a cross-section of the arm, we see that the base of the plates of the dorsal intermediate series is prolonged inwards (and ventralward), so that a single isolated plate appears club-shaped.

All the plates are pitted for granules except at the extreme margin.

The supero-marginal plates are from twelve to thirteen in number, the infero-marginals from thirteen to fourteen in number. The arm is very high and both infero- and supero-marginal plates appear in dorsal view. In fact, the infero-marginal plates do not take any part in the formation of the actinal surface. This is parallelled in modern species of *Pentaceros*, e. g. *P. clavatus*. In marginal view the supero- and infero-marginals at the extremity of the ray very distinctly alternate. This alternation persists at the base of the arms, but here it is not always so obvious.

The supero-marginals are much higher than the infero-marginals, and also more oblong in shape. Both supero- and infero-marginal plates are regularly but coarsely pitted for granules.

In specimen figured on Pl. XX, fig. 1, we obtain the following measurements:

Breadth of fifth infero-marginal from the extremity of ray . 5·8 mm.
Length „ „ „ „ „ . 4 „
Breadth „ supero-marginal „ „ „ . 9·2 „
Length „ „ „ „ „ . 3·9 „
Breadth „ radialia „ „ „ . 5·8 „
Length „ „ „ „ „ . 4·4 „
Width of ambulacral groove . . . . . 1·2 „

The ocular is visible in this specimen. It is about 1·6 mm. in length and breadth. The extremity is slightly pointed, and its ventral surface is hollowed out.

A ventral view is figured on Pl. XXI, fig. 1. Ventro-lateral plates extend almost to the extremity of the arms. These are, as usual, rather greater in breadth than in length. The adambulacral plates appear to be about half the length of the bordering actinal plates. Their armature consists of several rows of spinelets arranged in pairs.

A few intermarginalia are present in the interradii. They, as usual, press the supero- and infero-marginalia on to the abactinal and actinal surfaces of the disc respectively.

Locality and Stratigraphical Position.—Upper Chalk, Bromley, Kent; according to Dr. Rowe, probably from the Chislehurst caves near that locality.

Variations.—Variations occur amongst all the specimens, especially with regard to the ornamentation of the plates and the madreporite. The British Museum specimens, 48748, which occur together in a slab, are especially note-worthy, inasmuch as the lowest situated individual possesses on the disc no plate, which is bulbiform or raised conspicuously above the remainder.

2. Pentaceros Boysii, *Forbes*, sp.   Pl. XXII, figs. 4, 4 *a*, 4 *b*, 4 *c*; Pl. XXIII, figs. 1, 1 *a*, 1 *b*; Pl. XXVI, figs. 2, 2 *a*, 2 *b*.

Oreaster Boysii, *Forbes*, 1848.   Mem. Geol. Surv. Gt. Brit., vol. ii, p. 468.
—   —   *Forbes*, 1850.   In Dixon's Geology and Fossils of the Tertiary and Cretaceous Formations of Sussex, p. 328, pl. xxi, fig. 6.
—   —   *Dujardin et Hupé*, 1862.   Hist. Nat. Zooph. Échin. (Suites à Buffon), p. 389.
—   —   *Forbes*, 1878.   In Dixon's Geology of Sussex (new edition, Jones), pp. 362, 370, pl. xxi, fig. 6.

Specific Characters.—The primary radialia and interradialia are large hemispheroid punctate tubercles. R : r : : 80 mm. : 18 mm. Rays well produced, steep-sided, almost square in section, and tapering gradually to the extremity. Only a few of the plates of the disc enter the base of the arm. Supero- and infero-marginal plates adjunct, the intermarginalia being represented only by a few scattered granules.

Material.—The type specimen was said by Forbes (1848) to be in the collection of the Marquess of Northampton. The specimen figured and described in Dixon's 'Geology of Sussex' (see reference) was said by Forbes to have been " discovered by Major Boys and formed part of his interesting collection." This statement does not preclude the hypothesis that the specimen figured was also the type specimen. Neither specimen (if there were two) can now be traced. The following description is based chiefly on a specimen in the Sedgwick Museum, Cambridge (Pl. XXVI, fig. 2), which shows the actinal surface of the arms and a portion of the disc. It is supplemented by reference to a less nearly perfect specimen preserved in the British Museum of Natural History (J. Simmons' Coll., 46600), which presents views of isolated rays (Pl. XXII, fig. 4), and an isolated ray seen from the dorsal surface (Pl. XXIII, fig. 1) in the same museum (Dixon Coll., 48083).

Description.—The disc is covered with a number of rounded or irregularly-shaped plates. A circlet of large tubercles is very distinct and characteristic of the species. These tubercles are hemispherical and not so swollen as those of *P. bulbiferus*. They are smooth, and possess a fine distinct ornament, thus distinguishing them from the circlet of *P. coronatus*. Their diameter is about 8·5 mm., and they seem to be arranged radially and interradially, making a total of ten. The madreporite was figured by Forbes. It is roughly triangular in shape.

The arms are well produced. $R = 80$ mm. and $r = 18$ mm., the major radius being thus about four and a half times the minor radius. They taper gradually to the extremity. The breadth of the ray about the fourth supra-marginal plate is 6·8 mm. The height of the ray at the same spot is almost exactly the same. The rays are steep-sided, and consequently appear almost square in cross section.

The supero-marginalia are adjunct throughout almost the whole length of the ray, for only one or two single radialia enter the base of the ray. At the base of the ray they are flat and slightly rhomboidal. They possess an anterior indentation on their inner surface and are about 3·5 mm. in breadth. They gradually diminish in size distally and at the same time become distinctly swollen. They number about twenty-eight.

The infero-marginal plates are approximately of the same size and number as the supero-marginals. Both series imbricate slightly. The ornament of these plates consists of a number of fine granules in the centre, while there is a distinct margin without granulations.

Between the supero- and infero-marginal plates a few scattered granules represent a slight development of the intermarginalia.

The adambulacrals are a series of small oblong plates. They border the inferomarginals from about the eleventh supra-marginal onward. They are much worn, and but slight traces of their armature remain. About five adambulacrals occupy the same length as two infero-marginal plates. Proximally there is a single row of small plates which separate the two series.

Only a few scattered ossicles of the actinal surface of the disc remain.

Locality and Stratigraphical Position.—Upper Chalk, Kent.

Remarks.—Valette ('Bull. Soc. Yonne,' 1902) has described a number of species of starfishes from the Senonian of the South of France. The remains are found as scattered ossicles. Some of these are grouped by Valette as a new species which he calls *P. senonensis*. They are noticed by the author to resemble *P. Boysii* except that they are smooth and therefore do not have the ornament possessed by *P. Boysii*. Valette regards this absence of ornament as rendering them specifically distinct from *P. Boysii*, as other ossicles found in close proximity still possess the ornament. In view of the vagaries of the way in which solution may occur, I cannot admit this contention and consider that it is much more probable that the ossicles at one time possessed ornament and were identical with *P. Boysii*. All the other ossicles except those of the so-called *Arthraster senonensis* (*vide infra*, p. 92) were identified with English Cretaceous genera, which would support this contention.

3. PENTACEROS CORONATUS, *Forbes*, sp.   Pl. XIX, figs. 1, 1 *a*; Pl. XXIV, figs. 2,
2 *a*, 2 *b*, 2 *c*; Pl. XXV, fig. 9.

OREASTER CORONATUS, *Forbes*, 1848.   Mem. Geol. Surv. Gt. Brit., vol. ii, p. 467.

—   —   *Forbes*, 1850.   In Dixon's Geology and Fossils of the Tertiary and Cretaceous Formations of Sussex, pp. 327,
328, pl. xxi, fig. 7 *a—d*.

—   —   *Dujardin et Hupé*, 1862.   Hist. Nat. Zooph. Échin. (Suites
à Buffon), p. 389.

—   —   *Forbes*, 1878.   In Dixon's Geology of Sussex (new edition,
Jones), pp. 362, 370, pl. xxi, figs. 7, 7 *a—d*.

Specific Characters.—Disc large, with conspicuous nodular primary radialia and
interradialia.   The major radius is about five times the length of the minor radius.
Sides of arms very steep, so that the arm appears to be square in cross section.
A triple row of intermarginalia present in the interbrachial areas.

Material.—The type specimen of this species is preserved in the British Museum
of Natural History (Dixon's Coll., 35480).   Unfortunately, only one arm and a
portion of the disc are preserved.   A further specimen, registered E. 2562, from the
cabinet of Mrs. Smith, of Tunbridge Wells, is preserved in the same museum, and
another example is to be seen in the Museum of Practical Geology, Jermyn Street.

Description.—The most conspicuous feature of the disc is the circlet of ten
"large, more or less polygonal nodose pyramidal tubercles."[1]   These are the
primary radialia and interradialia.   The interradial tubercles are rather larger
than the radial tubercles, the former measuring 9·2 mm., the latter 7·7 mm. in
diameter.   The remainder of the disc is covered by irregularly shaped plates.

The madreporite has been broken away from the disc of the specimen no. 35480.
It is figured Plate XXV, fig. 9.

R : r : : 58 mm. : 19 + mm. in the type specimen where the single arm is
broken short.   In specimen no. E. 2562 R : r : : 100 mm. : 20 mm.   The arms
are 30 mm. broad at the base.   Their surface is flat, and the sides slope away
at right angles, so that a cross section of the arm is square.

Both radial and adradial plates are present in the base of the ray.   The
adradials are irregular in shape and soon disappear.   The radials are roughly
oblong in appearance, and exist throughout that portion of the arm preserved.
They diminish in size, however, distally.

The supero-marginal plates are indented on their anterior median surface.

[1] Forbes, in Dixon's 'Geology of Sussex,' p. 327.

They appear to imbricate slightly at their margins. The breadth of the fourth supero-marginal is 7 mm., the length 4 mm., and the height 3·5 mm. The height of the ray at this point is 12·2 mm.

The infero-marginal plates are opposite to the supero-marginals. They are approximately about the same size and number. Between the supero- and infero-marginal series a triple series of intermarginalia occurs in the interradial areas. The inner and larger intermarginals persist throughout the greater part of the length of the arm. It is this intercalated series which gives to the arm its great proportionate depth. The outer and smaller series disappear at about the seventh and ninth infero-marginal plates.

The ornamentation of the plates appears to have been worn away, although upon many of the plates a distinct marginal area may be seen.

Upon most of the plates there occur small entrenched pedicellariæ which are very characteristic of this species of *Pentaceros*. They consist of a small pit from which radiate two fine entrenchments (see Pl. XXIV, fig. 2 *a*).

One of the rows of specimen no. E. 2562 is distorted so as to bring the ventral surface into view. This shows that the ventro-lateral plates extend well towards, and perhaps all the way to, the extremities of the arm.

Locality and Stratigraphical Position.—The locality of the type specimen is given as Lower Chalk, Washington, Sussex. The specimen registered E. 2562 is from the Lower Chalk, Burham, Kent, and the specimen in the Museum of Practical Geology is from the Lower Chalk, Dover.

Remarks.—The specimen registered E. 2562 presents only one or two pedicellariæ, which are so characteristic and numerous on the other two specimens.

4. PENTACEROS SQUAMATUS, *Forbes*, sp. Pl. XXV, figs. 3, 3 *a*, 3 *b*, 3 *c*.

OREASTER SQUAMATUS, *Forbes*, 1848. Mem. Geol. Surv. Gt. Brit., vol. ii, p. 468.
— — *Forbes*, 1850. In Dixon's Geology and Fossils of the Tertiary and Cretaceous Formations of Sussex, p. 328, pl. xxiii, fig. 7.
— — *Dujardin et Hupé*, 1862. Hist. Nat. Zooph. Échin. (Suites à Buffon), p. 389.
— — *Forbes*, 1878. In Dixon's Geology of Sussex (new edition, Jones), pp. 363, 370, pl. xxiii, fig. 7.

Specific Characters. — Disc high, with conspicuous primary radialia, interradialia and centrale. Major radius about four times the length of the minor radius. Only radialia enter the base of the arm. Dorsal surfaces of arms flat, sides slope away at an obtuse angle from this. Ossicles distinctly imbricating. A few intermarginalia present.

Material.—The only specimen of this species is preserved in the Brighton Museum. The specimen consists of the disc and a portion of three arms. On the whole little displacement of the ossicles has taken place.

Description.—The disc is strongly convex, and is covered with the circlet of primary radialia and interradialia which are disposed around the centrale. All these ossicles appear shaped like a breast-plate. The centrale has a diameter of 4·2 mm. The primary interradialia are larger, possessing a diameter of 5·3 mm., whilst the primary radialia are the smallest of the series, measuring only 3·7 mm. across. Between the centrale and the primary interradialia a number of irregularly distributed plates appear. In the next right-hand interradius to the madreporite a number of these appear to have surrounded an anal opening. The primary interradialia almost touch one another, and the radialia consequently rest on the bases of pairs of ossicles. A few adradialia are present, but they are confined to the disc. A pair of them help to enclose the madreporite, which is a polygonal plate 9 mm. in greatest diameter. The ornamentation of the ossicles is rather coarse when present, but usually it is very much worn away.

The arms are well produced. R : r : : 30 + mm. : 7·8 mm. They are 1·3 mm. in breadth at the base. After the fourth or fifth radiale the remainder become minute but persist throughout the length of the arm preserved.

The supero-marginalia are finger-shaped; they, as also the infero-marginalia, distinctly imbricate. The dimensions of the third supero-marginal, reckoning from the median interradial line, are as follows : length 2·3 mm., breadth 3·1 mm. The long axes of the supero-marginal plates slope away distally, thus causing pairs of plates to assume the shape of arms of a **V**. They are at least thirteen in number.

The infero-marginal plates are similar in size and number to the supero-marginal series. In the interradii a few intermarginalia are present. These force the supero-marginal series to the surface of the disc.

Nothing is known of the ventral surface.

Locality and Stratigraphical Position.—Upper Chalk, Woolwich.

5. PENTACEROS OCELLATUS, *Forbes*, sp. Pl. XXV, figs. 4, 4 *a*.

OREASTER OCELLATUS, *Forbes*, 1848. Mem. Geol. Surv. Great Brit., vol. ii,
p. 468.
-- — *Forbes*, 1850. In Dixon's Geology and Fossils of the
Tertiary and Cretaceous Formations of Sussex,
p. 329, pl. xxi, fig. 13.
— — *Dujardin et Hupé*, 1862. Hist. Nat. Zooph. Echin.
(Suites à Buffon), p. 389.
— — *Forbes*, 1878. In Dixon's Geology of Sussex (new edition,
Jones), pp. 364, 370, pl. xxi, fig. 3.
PENTACEROS — *McPherson, W.*, 1902. Rep. Brighton Nat. Hist. Soc.

Specific Characters.—Ventro-lateral plates (as probably also the dorsal plates) depressed and finely striated on their truncated surface so as to simulate the surface of a madreporite, with sides rugged and ocellato-punctate. Between these plates smaller ossicles of a similar character are interspersed.

Material.—But one specimen of this species was known to Forbes. This is preserved in the British Museum of Natural History (Dixon Coll., E. 2571). It is a mass of ossicles which look as if they were derived from the dorsal surface of the disc. They are more spheroidal and somewhat larger than the ossicles of the ventral surface of the more nearly perfect example discovered by Mr. William McPherson in the Senonian Marsupites band at Brighton. This he presented to the British Museum (Natural History) in 1901 (E. 5012).

Description.—The disc and arms are unknown. The specimen no. E. 5012 shows a well-preserved portion of the ventral surface. The mouth-angles were occupied by single initial rhomboidal ossicles. To these succeed the ventro-lateral ossicles which border the ambulacral groove. These are pentagonal ossicles of very uniform size. The length of the exposed sides of the ossicles bordering the groove is 4·4 mm. and the greatest breadth of an ossicle 4·2 mm. The remaining ventro-lateral plates are hexagonal, but of almost the same dimensions, although the plates appear to become a little larger distally.

The plates overlap one another considerably, rendering precise measurement difficult. Between the larger plates are interspersed large numbers of smaller and more irregular ossicles which fill up the angles between their sides. The whole test would be thus very strongly built.

Both larger and smaller plates are curiously similar in appearance. The madreporiform striations on the truncated summits and the ocellato-punctate sides give a most characteristic appearance and render the species unmistakably distinct from all known species of *Pentaceros*.

The ambulacral groove is 3·5 mm. wide. The adambulacrals are difficult of recognition and have probably for the most part been lost, but a large number of the hour-glass shaped ambulacrals may be seen.

Locality and Stratigraphical Horizon.—Upper Chalk, Kent; Upper Senonian, Brighton.

6. Pentaceros abbreviatus, n. sp.   Pl. XXIV, figs. 1, 1 *a*, 1 *b*, 1 *c*.

Specific Characters.—Body of medium size. Arms moderately produced, but their breadth making them appear stumpy, rounded at the extremities, and hemispherical in cross section. Five series of dorsal ossicles enter their base. Of these the radialia and adradialia persist throughout the length of the arm. A few small intermarginalia are present.

Material.—There is only one specimen known of this species, and of this practically all that remains are two arms. It is preserved in the British Museum of Natural History (J. Tennant's Coll., 57538).

Description.—These arms are characteristically wide, the width of the arm at the base being 31 mm. They narrow very gradually towards the extremity. Throughout the ray the ossicles, except for the differences noted below, are very similar in appearance. At the base of the ray, where dorso-lateralia also enter into the composition of the dorsal skeleton, they are rounded and possess interspaces of considerable extent. These interspaces are often filled by smaller granules arranged irregularly. At times, however, between two radialia or adradialia one of the smaller ossicles is arranged in a very regular and alternating manner. Both large and small ossicles are finely granulated, and the large ossicles alone are perforated for pedicellariæ. The average size of the larger ossicles at the base of the arms is about 6 mm. Towards the extremity of the ray the radialia, adradialia, and marginalia become hexagonal, and fit very closely so as to make a compact skeleton. The terminal ocular plate is hexagonal and conspicuous. It has a flattened articulation which undoubtedly was originally occupied by a spine. Several of the other dorsal plates in the distal portion of the ray also possess similar articulation.

The supero-marginalia and infero-marginalia are equal in number. There were probably thirteen of each in the space between an interradius and an extremity of an arm.

In the interbrachial arc there is a series of minute granular intermarginalia.

The traces of the disc which are present suggest that the ossicles of this region were oval in shape and minute in size. I exposed a portion of the ventral surface of the arm, but, unfortunately, little trace of structure was shown. The ventro-lateralia extended to the extremity of the ray. The ridges of the adambulacral armature are lost.

Locality and Stratigraphical Position.—Upper Chalk, Charlton, Kent.

7. PENTACEROS BISPINOSUS, n. sp.   Pl. XXIII, figs. 3, 3 *a*, 3 *b*, 3 *c*.

Specific Characters.—Disc large. Arms moderately produced. Single isolated marginal ossicles vertebra-shaped with biconcave extremities. Ventro-lateral plates with strongly marked sockets for two or more spines.

Material.—The only specimen of this species is that preserved in the British Museum of Natural History (H. W. Taylor's Coll., 35482). Only the ventral surface is exposed, and this is very much distorted.

Description.—The disc appears to have been large. Its actinal surface is covered with a number of sub-equal oblong or polygonal plates, which possessed sockets in which fitted spines (Pl. XXIII, fig. 3*c*). These plates are 4·8 mm. long, and 3·1 mm. wide.

The arms are moderately broad, and at least four series of ventro-lateral plates enter at the base. R : r : : 60 mm. : 20 mm. (approximately), the major radius therefore measuring about three times the minor radius. The marginal ossicles are shaped very much like the centrum of a vertebra, and are biconcave. They possess a distinct granulation in their central region, which is surrounded by a wide margin. The infero-marginals at the base of the ray are about 3·2 mm. wide and 2·1 mm. long. There were probably sixteen of them from the interradius to the extremity of the ray.

The specimen is otherwise so distorted that little can be made of its structure.

Locality and Stratigraphical Position.—Upper Chalk, Sittingbourne, Kent.

The following are placed provisionally in the genus *Pentaceros* :

8. Pentaceros punctatus, n. sp.   Pl. XXVI, figs. 1, 1 *a*, 1 *b*.

Specific Characters.—Body of large size.   Marginal series of plates possessing well-developed foraminate pedicellariæ.   Intermarginalia present.

Material.—The only example of this species is a fragmentary portion of an arm preserved in the British Museum (Natural History) and bearing the registered number E. 2561.

Description.—The body of the starfish must have been of large size.   The supero-marginals are in contact in the extremity of the arm, although interspersed granular plates occur.   Proximally, at least, radial plates were present.   The largest supero-marginal present is 11·2 mm. high and 6·5 mm. long in its widest point.   It is of rather irregular shape and possesses two foraminate pedicellariæ. The infero-marginal plates alternate with the supero-marginals.   They are of the same height as the supero-marginals but only 5 mm. broad.   They are oblong in shape ; the two interior corners of the oblong, however, are cut away, making the ossicles six-sided.   The foramina once occupied by the pedicellariæ are deep and often situated in a depression.   From the foramen itself ridges may run out, which probably served for the attachment of muscles.

The infero-marginal series border only the side of the arm and take but little part in the formation of the ventral surface.

An intermarginal series of rounded granular plates occurs.

Locality and Stratigraphical Position.—Upper Chalk.

9. Pentaceros pistilliferus, *Forbes* sp.   Pl. XXV, fig. 5.

Oreaster pistilliferus, *Forbes*, 1848.   Mem. Geol. Surv. Great Brit., vol. ii, p. 467.
—   —   *Forbes*, 1850.   In Dixon's Geology and Fossils of the Tertiary and Cretaceous Formations of Sussex, p. 329, pl. xxi, fig. 15.
—   pistilliformis, *Dujardin et Hupé*, 1862.   Hist. Nat. Zooph. Échin. (Suites à Buffon), p. 389.
—   pistilliferus,   *Forbes*, 1878.   In Dixon's Geology of Sussex (new edition, Jones), pp. 363, 370, pl. xxi, fig. 15.

Specific Characters.—Primary radialia (or interradialia) large with a dilated summit which possesses no ornament and is excavated into pits. " Ossicles of the arm are narrow, shuttle-shaped, tumid in the centre and slightly impressed towards each extremity " (Forbes).

Material.—Several fragmentary remains of this species are known. The most nearly perfect remains are those in the Museum of Practical Geology, Jermyn Street. Other specimens are in the British Museum (Natural History), registered E. 5037, 57624, E. 2564 (all Pl. XXV, fig. 5), 7600, E. 25637, E. 2565.

Description.—Nothing is known further than the description given in the diagnosis. Forbes' description reads as if he had described the species from the specimen in the Museum of Practical Geology. This originally was in the collection of the Marquis of Northampton. Forbes seems to have described the large ossicles upside down. He also says they were in a circlet of five, which is not apparent in any specimen known. The roughened and pitted surface recalls in some respects the primary radialia and interradialia of *P. coronatus*.

Locality and Stratigraphical Position.—Upper Chalk, Kent and Sussex.

10. PENTACEROS, sp.   Pl. XXV, fig. 7.

This specimen is preserved in the British Museum of Natural History (no. 5514).

It consists of five marginal plates which are 12·3 mm. high, and have an average length of 5·5 mm. The plates are rugged in appearance and the ornament is worn away. A cirral of a crinoid (probably *Bourguetiocrinus*) has become fixed between two of these plates.

11. PENTACEROS, sp.   Pl. XXV, fig. 8.

The only specimen is preserved in the Brighton Museum. It consists of a few marginal plates. The supero-marginals are rather irregular in shape, some being almost wedge-shaped. On an average they are 4 mm. high and 3·2 mm. long. The infero-marginals are opposite and equal in length to the supero-marginals. They are only 2·9 mm. high. The plates possess a distinct margin, but the ornament otherwise is worn away. There are a few small granular inter-marginalia.

With these plates is associated a large plate which appears to be a worn radial or interradial of *P. Boysii*.

Family—ASTROPECTINIDÆ (*Gray*, 1840), *emend.* Sladen, 1886.

Phanerozonate Asteroids with large marginal plates bearing spines or spiniform papillæ. Abactinal skeleton with true columnar papillæ. Actinal interradial areas small, interradial plates when present spinose. Ambulacral plates short and more or less compressed. Superambulacral plates present. Aproctuchous. Pedicellariæ rarely present.

Genus—ASTROPECTEN, *C. F. Schulze*, 1760.

Adambulacral plates touching the infero-marginal plates along the ray. Marginal and adambulacral plate not correspondent in length and number. Supero-marginal plates more or less well developed. Marginal plates long and more or less quadrate. Superior and inferior series subequal.

ASTROPECTEN, sp.   Pl. XXV, figs. 2, 2 *a*.

Material.—There is one specimen in the Sedgwick Museum at Cambridge, which looks like an *Astropecten*. It is figured on Pl. XXV, figs. 2 and 2 *a*. Practically only the marginal plates are preserved.

Description.—R : r : : 45 mm. : 15 mm. The interbrachial arcs are well rounded. The supero-marginalia are remarkably uniform in size throughout the greater portion of the ray. Their breadth is 4 mm. and length 1·7 mm. About thirty of these are present from the interradius to the extremity. At the apex of the ray these plates are adjunct. The upper surface of each plate is rounded.

The infero-marginalia are equal in size, opposite to, and, as far as one can judge, similar in appearance to, the superior series. There is a distinct groove between the two series.

Locality and Stratigraphical Position.—Upper Greensand, Blackdown (?).

PLATE XVII.

METOPASTER MANTELLI, *Forbes*, sp. (Page 38.)

From the Upper Chalk.

Fig.
1. Actinal aspect; natural size. (Coll. Brit. Mus., 40402.)
   a. Infero-marginal plate; magnified 4 diameters.

MITRASTER COMPACTUS, *Forbes*, sp. (Page 67.)

From the Upper Chalk.

2. Actinal aspect; copied from Forbes in Dixon's 'Geology of Sussex,' pl. XXII, fig. 3.

COMPTONIA COMPTONI, *Forbes*, sp. (Page 69.)

From the Upper Greensand.

3. Actinal aspect; natural size. (Coll. Brit. Mus., 34311.)
   a. Infero-marginal plate; magnified 3 diameters.
   b. Lateral view of interbrachial arc; natural size.

COMPTONIA ELEGANS, *Gray*. (Page 71.)

From the Upper Greensand.

4. Abactinal view; natural size. (Coll. Brit. Mus., E. 2567.)
   a. Actinal view of same specimen; natural size.

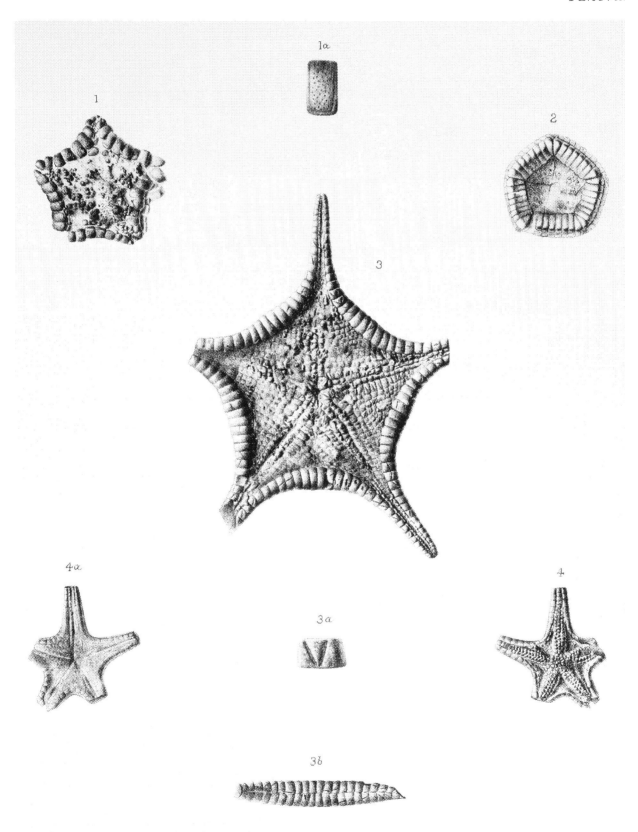

A.H.Searle del. et lith.

Pitcher L.^{td} imp.

CRETACEOUS ASTEROIDEA .

PLATE XVIII.

Arthraster Dixoni, *Forbes.*　(Page 91.)

From the Lower Chalk.

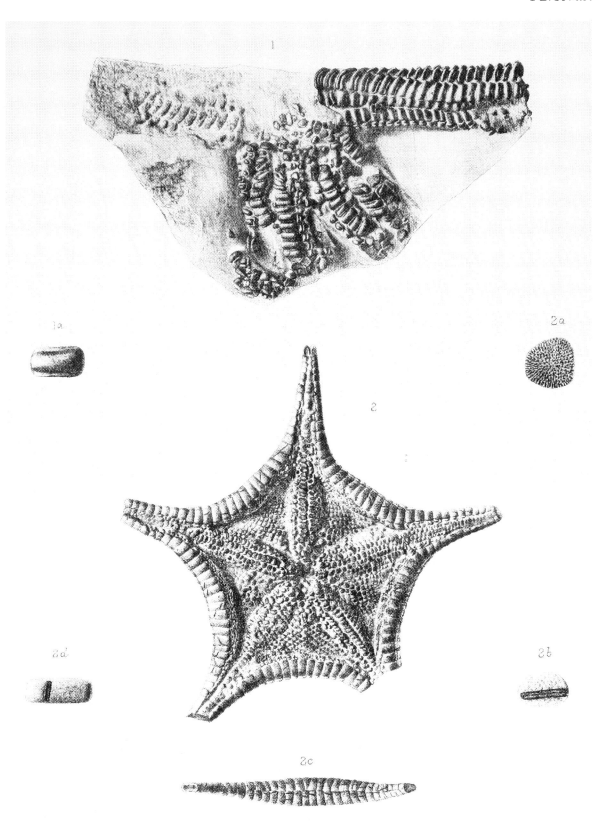

A.H.Searle del. et lith.

CRETACEOUS ASTEROIDEA.

Pitcher Lᵗᵈ imp.

PLATE XIX.

PENTACEROS CORONATUS, *Forbes*, sp.   (Page 82).

From the Lower Chalk.

Fig.

1.   Abactinal aspect; natural size.   (Coll. Brit. Mus., E. 2562.)
 a. Lateral view of arm ; natural size.

METOPASTER PARKINSONI, *Forbes*, sp.   (Page 31.)

From the Upper Chalk.

2.   Actinal aspect ; natural size.   (Coll. Brit. Mus., E. 5027.)
 a. Lateral view ; natural size.
 b. Ventro-lateral plate ; magnified 5 diameters, showing entrenched pedicellaria.
 c. Supero-marginal plate ; magnified 3 diameters.

NYMPHASTER COOMBII, *Forbes*, sp.   (Page 15.)

From the Upper Greensand.

3.   Actinal aspect ; natural size.   (Coll. Brit. Mus., 48620.)

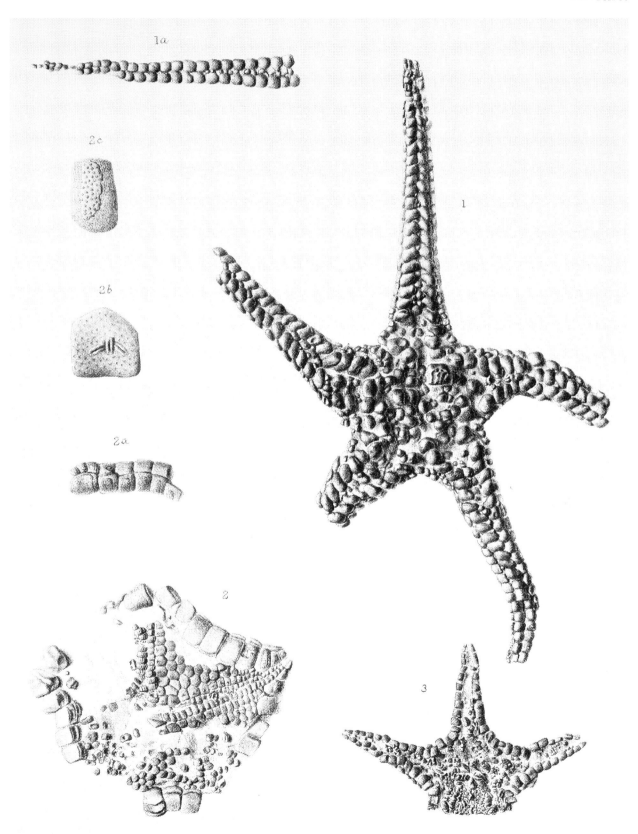

A.H.Searle del. et lith.

Pitcher L^{td} imp.

CRETACEOUS ASTEROIDEA .

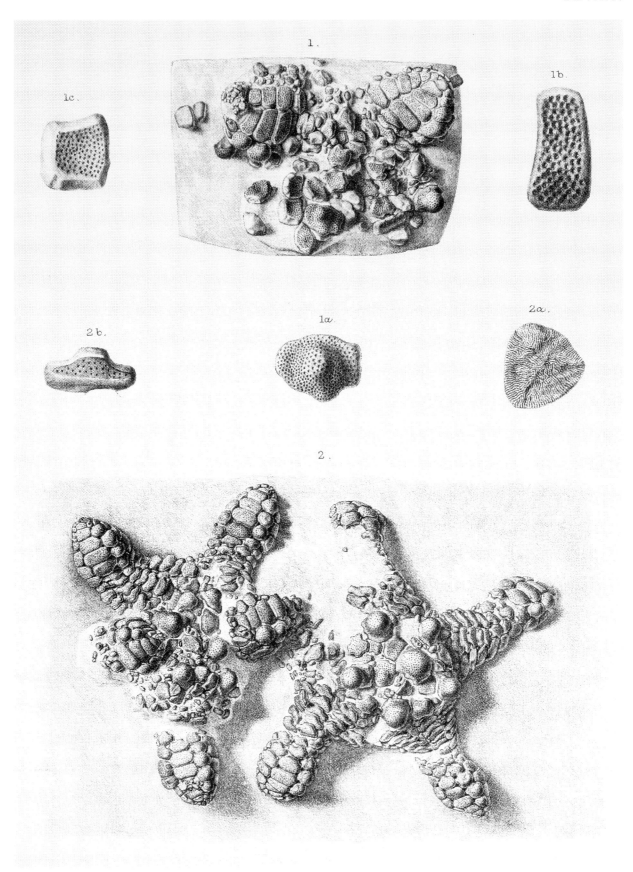

A.H.Searle del. et lith.

Pitcher L.^{td} imp.

CRETACEOUS ASTEROIDEA.

PLATE XXI.

Pentaceros bulbiferus, *Forbes*, sp.   (Page 77.)

From the Upper Chalk.

Fig.

1.   View of the extremities of three arms and portion of disc; natural size. (Coll. Brit. Mus., E. 5040.)

  a. Ventro-lateral plate; magnified 6 diameters.

  b. Actinal view of the extremities of two of the above arms; natural size.

Pentagonaster robustus, n. sp.   (Page 73.)

From the Upper Chalk.

2.   Abactinal aspect; natural size.   (Coll. Brit. Mus., 48085.)

  a. Lateral view of an arm; natural size.

Pentaceros bulbiferus, *Forbes*, sp.   (Page 77.)

From the Upper Chalk.

3.   Abactinal aspect; natural size.   (Coll. Brit. Mus., E. 5041.)

  a. Madreporite; magnified 6 diameters.

4.   Abactinal aspect; natural size.   (Coll. Brit. Mus., 40399.)

  a. Madreporite; magnified 6 diameters.

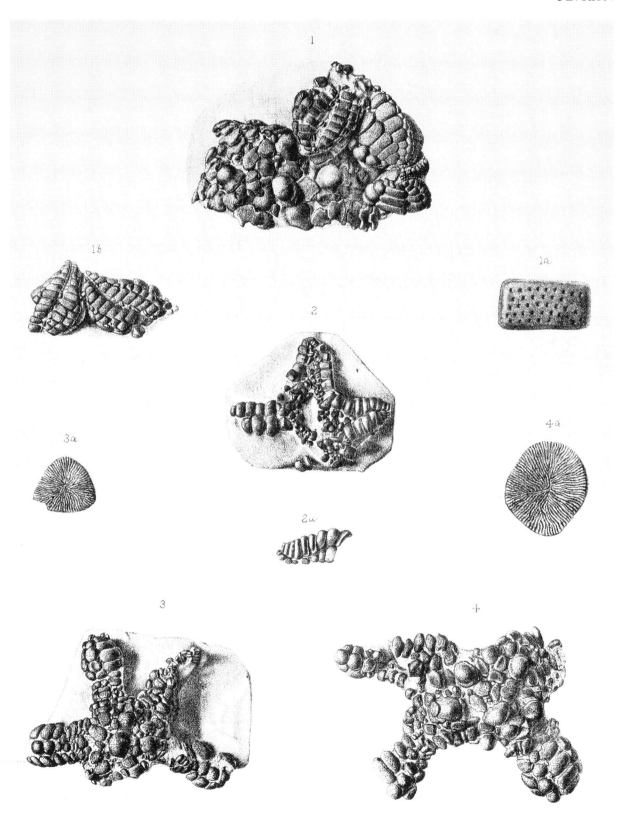

A.H.Searle del. et lith.

Pitcher L^{td} imp.

CRETACEOUS ASTEROIDEA.

PLATE XXII.

PENTAGONASTER OBTUSUS, *Forbes*, sp.   (Page 74.)

From the Upper Chalk.

FIG.

1. Abactinal aspect; natural size.   (Coll. Brit. Mus., 40400.)

  a. Madreporite; magnified 10 diameters.

  b. Lateral view of margin, abactinal side upwards; natural size.

2. Actinal aspect of two specimens; natural size.   (Coll. Brit. Mus., 35481.)

  a. Lateral view of extremity of arm of the underlying specimen; natural size.

3 *a.* Dorsal view of extremity of arm; natural size.   (Coll. Brit. Mus., E. 5039.)

  b. End view of extremity of arm; natural size.   (Coll. Brit. Mus., E. 5038.)

  c. Side view of extremity of same arm; natural size.

  d. Ventral end of extremity of same arm; natural size.

  e. Supero-marginal plate; magnified 3 diameters.

  f. Infero-marginal plate; magnified 3 diameters.

  g. Adambulacral plate; magnified 5 diameters.

PENTACEROS BOYSII, *Forbes*, sp.   (Page 80.)

From the Upper Chalk.

4. Abactinal view; natural size.   (Coll. Brit. Mus., 46600.)

  a. Isolated ossicle of disc; magnified 4 diameters.

  b. Isolated ossicle of disc; magnified 4 diameters.

  c. Lateral view of arm; natural size.

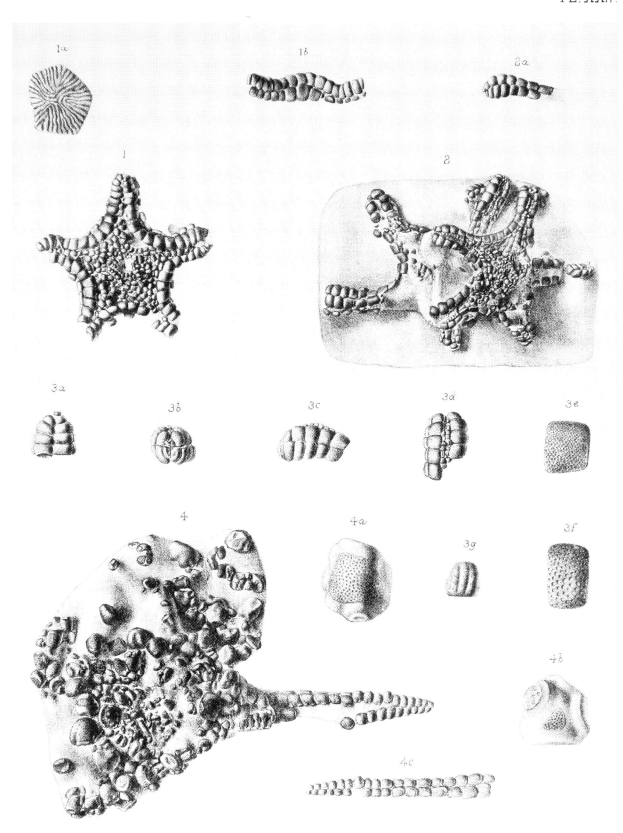

A.H.Searle del.et lith.

Pitcher L^td imp.

CRETACEOUS ASTEROIDEA.

PLATE XXIII.

Pentaceros Boysii, *Forbes*, sp.   (Page 80.)

From the Upper Chalk.

Fig.

1.   Supero-marginal plate ; magnified 4 diameters.   (Coll. Brit. Mus., 48083.)
   a. Lateral view of extremity of arm from which the above supero-marginal plate was drawn ; natural size.
   b. Dorsal view of extremity of same arm ; natural size.

Pentaceros bulbiferus, *Forbes*, sp.   (Page 77.)

From the Upper Chalk.

2.   Abactinal aspect ; natural size.   (Coll. Brit. Mus., E. 5042.)
   a. Madreporite ; magnified 5 diameters.

Pentaceros bispinosus, n. sp.   (Page 87.)

From the Upper Chalk.

3.   Actinal aspect ; natural size.   (Coll. Brit. Mus., 35482.)
   a. Actinal view of extremity of arm ; natural size.
   b. Infero-marginal plate ; magnified 3 diameters.
   c. Ventro-lateral plate ; magnified 4 diameters.

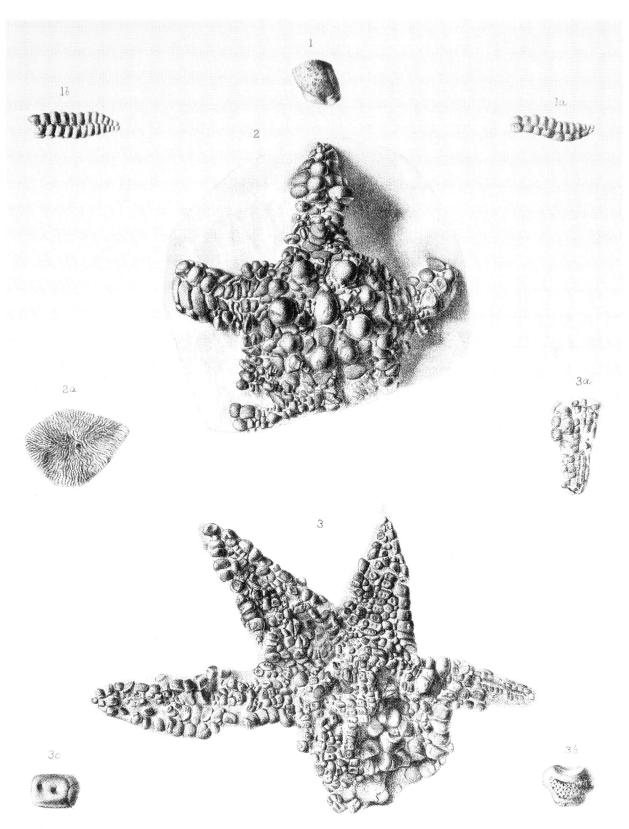

A.H.Searle del. et lith.

Pitcher L.^{td} imp.

CRETACEOUS ASTEROIDEA.

PLATE XXIV.

Pentaceros abbreviatus, n. sp. (Page 86.)

From the Upper Chalk.

Fig.

1. Abactinal aspect; natural size. (Coll. Brit. Mus., 57538.)
   a. Lateral view of arm; natural size.
   b. Radial plate; magnified 3 diameters.
   c. Two succeeding supero-marginal plates; magnified 3 diameters.

Pentaceros coronatus, *Forbes*, sp. (Page 82.)

From the Lower Chalk.

2. Abactinal aspect; natural size. (Coll. Brit. Mus., 35480.)
   a. Plate of disc; magnified 3 diameters.
   b. Supero-marginal plate; magnified 3 diameters.
   c. Lateral view of arm; natural size.
   [See also Pl. XXV, fig. 9.]

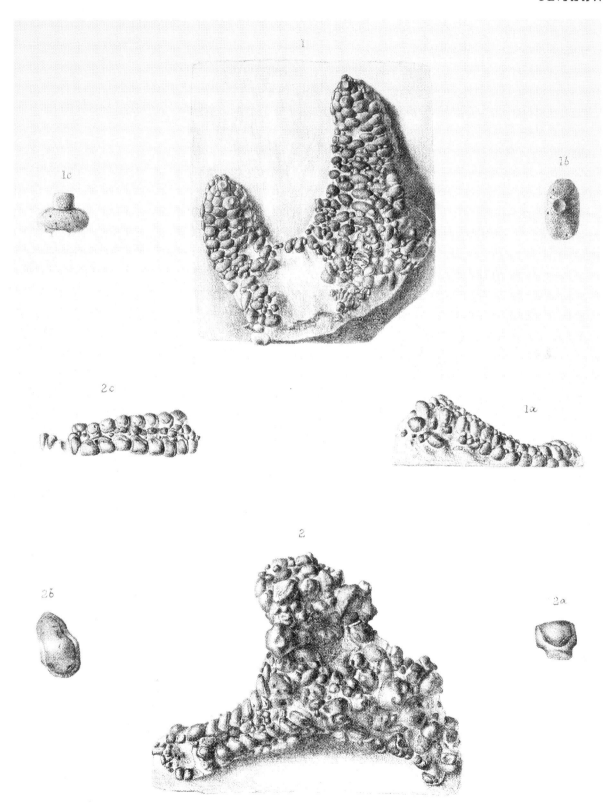

A.H.Searle del. et lith.

Pitcher L^{td} imp.

CRETACEOUS ASTEROIDEA .

PLATE XXV.

NYMPHASTER RADIATUS, n. sp. (Page 73.)

From the Lower Chalk.

FIG.
1. Abactinal view; natural size. (Coll. Brit. Mus., E. 375.)
  a. Lateral view of margin; natural size.
  b. Supero-marginal ossicle; magnified 4 diameters.

ASTROPECTEN ? n. sp. (Page 90.)

From the Upper Greensand.

2. Abactinal view; natural size. (Coll. Sedgwick Museum, Cambridge.)
  a. Isolated marginal ossicle; magnified 4 diameters.

PENTACEROS SQUAMATUS, *Forbes*, sp. (Page 83.)

From the Upper Chalk.

3. Abactinal view; natural size. (Willett Coll., Brighton Mus.)
  a. Madreporite; magnified 6 diameters.
  b. Marginal view of arm; natural size.
  c. Supero-marginal ossicle; magnified 6 diameters.

PENTACEROS OCELLATUS, *Forbes*, sp. (Page 85.)

From the Upper Chalk.

4. Actinal view; natural size. (Coll. Brit. Mus., E. 5012).
  a. Ventro-lateral ossicle; magnified 4 diameters.

PENTACEROS PISTILLIFERUS, *Forbes*, sp. (Page 88.)

From the Upper Chalk.

5. Ossicles of disc; natural size. (Coll. Brit. Mus.; from left to right the register numbers are E. 5037, 57634, E. 2564.)

GENUS ? sp. ? (Page 93.)

From the Chalk.

6. Ossicles; natural size. (Coll. Brit. Mus.)
  a. Isolated ossicle; magnified 4 diameters.

PENTACEROS ? n. sp. (Page 89.)

From the Chalk.

7. Marginal view of ossicles of arm; natural size. (Coll. Brit. Mus., 5514.)

PENTACEROS ? n. sp. (Page 89.)

From the Chalk.

8. Marginal view of ossicles; natural size. (Willett Coll., Brighton Mus.)

PENTACEROS CORONATUS, *Forbes* sp. (Page 82.)

9. Madreporite; magnified 4 diameters. (Coll. Brit. Mus., 35480.)

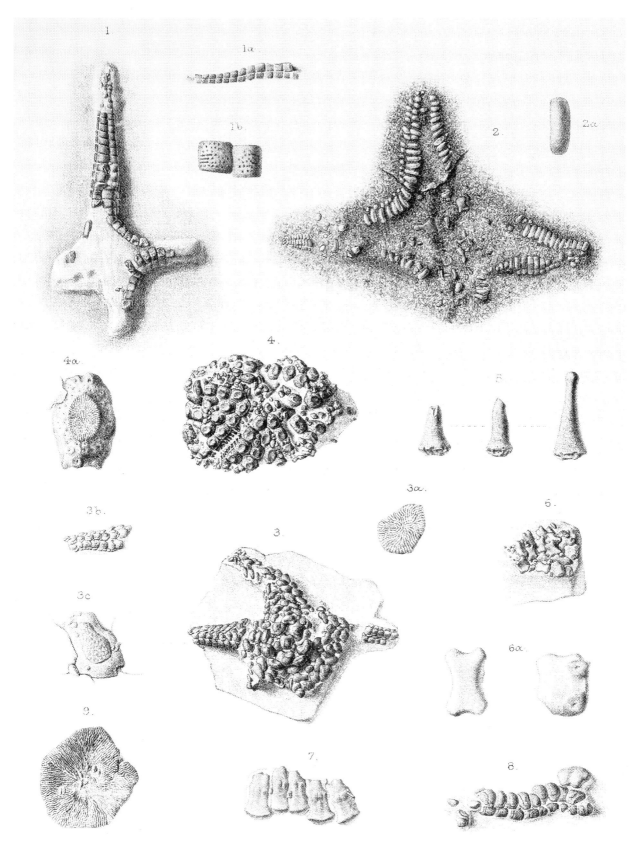

A.H.Searle del. et lith.

Pitcher Lᵗᵈ imp.

CRETACEOUS ASTEROIDEA.

PLATE XXVI.

Pentaceros punctatus, n. sp.　(Page 88.)

From the Upper Chalk.

Fig.
1.　Abactinal view of remains of arms; natural size.　(Coll. Brit. Mus., E. 2561.)
　a. Lateral view of margin; natural size.
　b. Enlarged view of single ossicle; magnified 2 diameters.

Pentaceros Boysii, *Forbes,* sp.　(Page 80.)

From the Upper Chalk.

2.　Actinal view; natural size.　(Coll. Sedgwick Mus., Cambridge.)
　a. Lateral view of margin; natural size.
　b. View of isolated ossicle; magnified 4 diameters.

Mitraster compactus, *Forbes,* sp.　(Page 67.)

From the Upper Chalk.

3.　Abactinal view; natural size.　(Willett Coll., Brighton Mus.)
　a. View of end of arm; magnified 4 diameters.
　b. Lateral view of supero-marginal ossicles; magnified 4 diameters.
　c. Lateral view of infero-marginal ossicles; magnified 4 diameters.

Calliderma mosaicum, *Forbes,* sp.　(Page 9.)

From the Upper Chalk.

4.　Actinal view of ambulacral groove; magnified 4 diameters.　(Coll. Sedgwick Museum, Cambridge.)
　a. Actinal view; natural size.
　b. Infero-marginal ossicle; magnified 4 diameters.

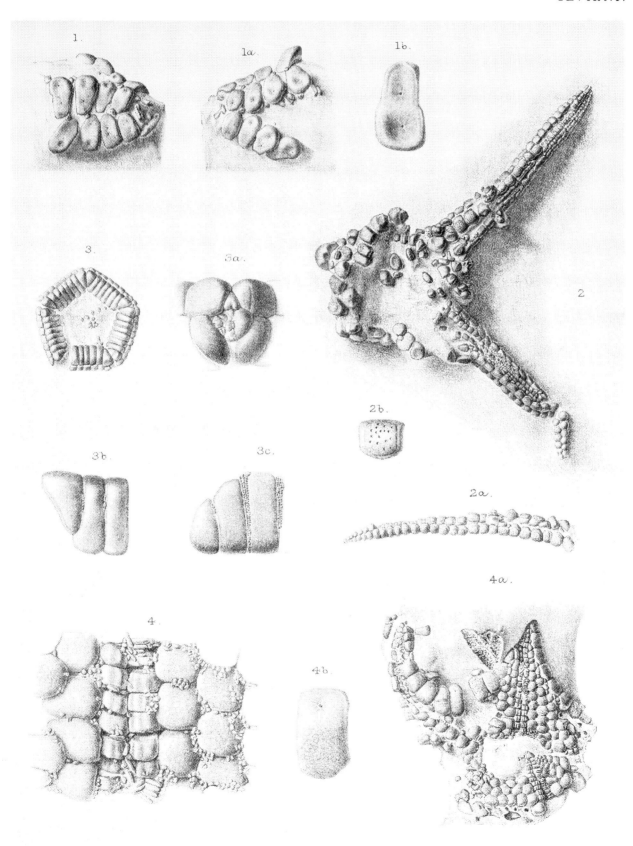

A.H.Searle del. et lith.

Pitcher L^{td} imp.

CRETACEOUS ASTEROIDEA.

Palæontographical Society, 1907.

A MONOGRAPH

ON THE

BRITISH FOSSIL

ECHINODERMATA

FROM

THE CRETACEOUS FORMATIONS.

VOLUME SECOND.
THE ASTEROIDEA AND OPHIUROIDEA.

BY

W. K. SPENCER, B.A., F.G.S.

PART FOURTH.

PAGES 91—132; PLATES XXVII—XXIX.

LONDON:

PRINTED FOR THE PALÆONTOGRAPHICAL SOCIETY.

1907.

PRINTED BY ADLARD AND SON, LONDON AND DORKING.

FAMILY UNCERTAIN.

Genus—ARTHRASTER, *Forbes*, 1848.

Arms stout and long. Radialia, marginalia, and ventro-lateralia form an alternating series of seven very completely articulating similar ossicles, which fit so closely as to leave no conspicuous interstices. Each ossicle consists of an oblong and flattened base with a surmounting ridge. Ventro-lateral plates on actinal surface of disc small and mammiform. Ossicles on abactinal surface of disc hemispheroid with a crenulated edge. All the ossicles possess, as ornament, hemispherical granular prominences.

1. ARTHRASTER DIXONI, *Forbes.* Pl. XVIII, figs. 1 and 1 *a*; Pl. XXIX, figs. 11 and 11 *a*.

ARTHRASTER DIXONI, *Forbes*, 1848. Mem. Geol. Surv. Gt. Brit., vol. ii, p. 467.

— — *Forbes*, 1850. In Dixon's Geology of Sussex, p. 336, pl. xxiii, fig. 1.

— — *Dujardin et Hupé*, 1862. Hist. Nat. Zooph. Échin. (Suites à Buffon), p. 437.

— — *Forbes*, 1878. In Dixon's Geology of Sussex (new ed., Jones), pp. 369 and 370, pl. xxiii, fig. 1.

Specific Characters.—Dorsal ridge of all of the arm ossicles well rounded. No spines present except on the adambulacral plates.

Material.—The best example of this very peculiar starfish is preserved in the British Museum of Natural History, Dixon Coll., 47000. The specimen consists of the remains of four arms, only one of which is at all well preserved. It is the type described by Forbes, and is figured in this Monograph on Pl. XVIII. A well-preserved fragment of an arm is also in the possession of Dr. Rowe, of Margate. I have referred two fragmentary specimens presented to the British Museum (E 5023 and E 5024) by Mr. W. McPherson, F.G.S., to this species (*vide infra*).

Description.—A section of the arm is similar at all points. It shows seven ossicles, namely, a radial, the pairs of supero- and infero-marginalia, and a pair of ventro-lateralia. These ossicles all alternate in series, and they closely fit the corresponding neighbouring plates in their respective series. The edges of the plates possess articulations which assist in forming this close union. All the plates

14

are generally similar in appearance. They differ, however, in measurements as detailed below.

Breadth of radialia	. 8·2 mm.
Length ,, ,, . . .	. 3·5 ,,
Breadth of supero-marginalia . .	. 6·2 ,,
Length ,, ,, . .	. 3·5 ,,
Breadth of infero-marginalia . .	. 5·9 ,,
Length ,, ,, . .	. 3·5 ,,
Breadth of ventro-lateralia . .	. 4·5 ,,
Length ,, ,, . .	. 3·5 ,,

It will thus be seen that the breadth of the plates diminishes as we proceed ventralwards. The ridges, also, on the plates, become more rounded in the same direction.

The ornament of the plates consists of hemispherical granular prominences of moderate size. They appear to have been especially prominent at the base of the ridge. No spine-pits are present.

The height of the ray is 16·5 mm., and the breadth is about the same. Post-mortem contraction has brought the ventro-lateralia of opposite sides into close approximation, in some cases totally obliterating the ambulacral groove. Along one or two of the arms some of the adambulacrals are still visible. They are 1·8 mm. broad and 1·2 mm. long. The portion of the plate nearest the ambulacral groove is depressed, giving the plate a two-storied appearance.

A few robust, rhomboidal, smallish ventro-lateralia are present at the base of the arm. Their breadth is 1·8 mm. They are mammiform. A few similar plates also enter the base of the arm.

The two collections of isolated ossicles presented by Mr. McPherson referred to above are very interesting. Each specimen consists of a single ossicle simulating one of the abactinal bulbiform ossicles of *Pentaceros*, but possessing the distinct *Arthraster* ornament, associated with plates which exactly match the ventro-lateralia of *A. Dixoni* and other plates which resemble the arm plates of this species except that the surmounting ridge is not so high. There is no doubt that the plates are those of a species of *Arthraster*. I have little hesitation, in spite of their occurrence in the Upper Chalk, in referring the ossicles to *A. Dixoni*, especially as it is a matter of common experience that species of Chalk starfish have a wide stratigraphical range.

Locality and Stratigraphical Position.—Forbes' type is from the Lower Chalk, Balcombe, Sussex. The specimen in the possession of Dr. Rowe was collected in the zone of *Terebratulina gracilis* in Devon. The specimens presented to the British Museum by Mr. McPherson are from the *Marsupites* zone, Brighton.

Remarks.—Forbes compared the genus *Arthraster* with the modern genus *Ophidiaster.* The larger amount of material known since that time does not allow us to recognise such affinity.

Valette (see above, p. 81) has described certain isolated plates, which are similar in form and size, as belonging to the genus *Arthraster*, and has called the species *A. senonensis.* These plates are smooth and show no trace of the surmounting longitudinal ridge or ornament which is so characteristic of *Arthraster.* Dom Aurélien Valette kindly lent these plates to Dr. Bather in order that I might examine them. I am therefore enabled to state that the plates are those of *Pycinaster angustatus.*

2. ARTHRASTER CRISTATUS, n. sp. Pl. XXIX, figs. 10, 10 *a*, 10 *b*.

Specific Characters.—Ridges of the radialia and supero-marginalia cristate. Upper surface of the ridge of all arm-ossicles possessing lipped pits formerly occupied by small spines.

Material.—The specimen figured on Pl. XXIX was restored by Dr. Blackmore, of Salisbury, from a number of isolated ossicles in his collection which were found in a single mass of chalk. These ossicles are the only material known of the species.

Description.—The dimensions of the ossicles are as follow:

Breadth of radialia . . .	5·4 mm.
Length ,, ,, . . .	3·2 ,,
Breadth of supero-marginalia . .	4·8 ,,
Length ,, ,, . .	3·2 ,,
Breadth of infero-marginalia . .	4·2 ,,
Length ,, ,, . .	3·2 ,,
Breadth of ventro-lateralia . .	3·8 ,,
Length ,, ,,	3·2 ,,

Just as in *A. Dixoni* the breadth of the plates diminishes ventralwards, and the ridges on the plates become more rounded in the same direction. The cristate ridges of the more dorsal plates are, however, very characteristic of the species, as are also the lipped pits on the summit of the ridges. The pits were formerly occupied by small spines. The base of the ridges of the plates possesses the granular hemispherical prominences, such as are also· met with in *A. Dixoni* and characterise the genus.

Locality and Stratigraphical Position.—Micheldever, Hants. Zone of *Micraster cor-anguinum.*

ADDENDUM (to *Phanerozonate Asteroids*).

Further investigation, as a result of the privilege of investigating the fine collection of Chalk Asteroids in the possession of Dr. Blackmore, of Salisbury, has enabled me to describe several new species belonging to genera which have been dealt with in previous pages. Some of these species had been recognised but not described by Dr. Blackmore, to whom I am indebted for very many valuable suggestions.

Family—PENTAGONASTERIDÆ, *Perrier*, 1884. (See p. 3.)

Genus—NYMPHASTER, *Sladen*, 1885. (See p. 14.)

5. NYMPHASTER RUGOSUS, n. sp. Pl. XXIX, figs. 7, 7 *a*.

Specific Characters.—All marginalia covered with granular prominences, which are closely crowded, and in no case arranged in a linear series. No spine-pits on the marginalia. Margin of disc lunate.

Material.—Two specimens are known of this species. They are preserved in the British Museum (Nat. Hist.), and bear the registered numbers 57516 (purchased of W. Griffiths) and 76002 (Capron Coll.). Both specimens are imperfect. The first-named specimen is figured Pl. XXIX, fig. 7, and is taken as the type.

Description.—The two specimens show the disc to have been small. The minor radius in the specimen 57516 measures about 11 mm. The arms are broken off short in both specimens, and therefore it is not possible to give the major radius.

There are about eight infero-marginalia in each interbrachial arc. These are all approximately equal in size, being 2·7 mm. long and 1·8 mm. broad. In shape they are oblong.

The margin of the disc is lunate. It is this character and the character of the granular prominences which distinguish the species from *N. radiatus*.

Locality and Stratigraphical Position.—Lower Chalk, Dover and Folkestone.

Genus—PYCINASTER,[1] nom. nov.

PYCNASTER, *Sladen*, 1891 (see p. 21), non *Pomel*, 1883. Classif. méthod. Echin., p. 42.

1. PYCINASTER ANGUSTATUS, *Sladen* sp. (see p. 21). Pl. IX, figs. 1 *a*, 1 *b*; Pl. XXI, figs. 2, 2*a*; Pl. XXV, fig. 7; Pl. XXVI, figs. 4, 4 *a*, 4 *b*.

This species appears to be quite common in the Upper Chalk. An exceedingly well-preserved specimen is in the collection of Dr. Blackmore, of Salisbury.

The following specimens, which belong to the genus *Pycinaster*, and probably to this species, have been erroneously ascribed by me to other genera and species in Part III of this volume (pp. 67–90).

The single specimen described on p. 73 as a new species, *Pentagonaster robustus*, is probably an immature form of this species. A collection of five ossicles described (p. 89) as *Pentaceros*, sp., and the specimen figured on Pl. XXVI as *Calliderma mosaicum*, also belong to the species. The latter specimen should be described as from the Upper Chalk.

Dom Aurélien Valette has courteously enabled me to examine the syn-types of his *Arthraster senonensis* (' Bull. Soc. Sci. Yonne,' 1902, p. 23). They prove to be marginals of *Pycinaster angustatus*. Four ossicles referred by him to his *Pentaceros senonensis* (*vide infra*) also belong to the present species.

2. PYCINASTER SENONENSIS, *Valette*, sp.   Pl. XXVI, figs. 1, 1 *a*, 1 *b*; Pl. XXIX, figs. 6, 6 *a*.

PENTACEROS SENONENSIS, *Valette*, 1902. Bull. Soc. Sci. Yonne, vol. lvi, pp. 17, 18, figs. 1, 2 (non 3—7).
— PUNCTATUS, *Spencer*, 1905. Antea, p. 88.

Dr. Blackmore's material enables me definitely to ascribe this species to *Pycinaster*, and to add the following new diagnosis and details :

Specific Characters.—Body of large size. Breadth of marginalia more than twice their thickness. All marginalia smooth or with very shallow hexagonal spine-pits.

Description.—The marginalia may be as much as 20 mm. high. They appear to be distinguished from the marginalia of *P. angustatus*, not only by their

[1] Πυκινός, compact, Homeric form of πυκνός. Dom Aurélien Valette kindly pointed out the prior use of *Pycnaster* to Dr. F. A. Bather, who suggests the above emendation.

magnitude, but also by the manner in which the upper surface is turned over so as to make the ossicle ſ-shaped (compare Text-fig. 23, p. 119).

Associated with these marginalia are found rounded, smooth ossicles, which are correspondingly large, being as much as 12 mm. in diameter, and which are undoubtedly ossicles from the abactinal surface of the disc. Their size and form (see Pl. XXIX, fig. 6) render them liable to be mistaken for ossicles of *Staurander-aster* (see p. 125). They do not, however, possess spine-pits, and species of *Stauranderaster* which do not possess spine-pits are very distinct, having nodular abactinal ossicles of a very characteristic appearance (compare *S. coronatus*, Pl. XXIV, fig. 2). There appears therefore to be no doubt that these ossicles should be referred to the genus *Pycinaster*, and probably to *P. punctatus*. The base of the abactinal ossicles of *P. angustatus* is produced, as also in *P. crassus* (Pl. XXIX, fig. 4 *a*), and quite different from the flattened base of these ossicles.

Remarks.—It might be urged that the differences which separate these ossicles from those ascribed to *P. angustatus*, are not sufficient warrant for a new species. I regard the differences, however, given above as important, and though several well-preserved specimens of *P. angustatus* are known, none approaches the large size which *P. punctatus* must have attained. On p. 81, *Pentaceros senonensis*, Valette, was regarded as probably identical with *P. Boysii*. Examination of the original specimens, which I owe to the courtesy of Dom Aurélien, shows that they belong to four different species : *Stauranderaster coronatus*, *S. argus*, *Pycinaster angustatus*, and my "*Pentaceros punctatus*." The last species is represented by two dorsal ossicles from Les Clérimois (figs. 1 and 2). With the concurrence of Dom Aurélien, I therefore take the original of his fig. 1 as type.

Locality and Stratigraphical Position.—The specimens in the possession of Dr. Blackmore are from East Harnham, Wilts., zone of *Actinocamax quadratus*.

3. Pycinaster crassus, n. sp. Pl. XXIX, figs. 1, 2, 2 *a*, 3, 3 *a*, 4, 4 *a*, 5.

Specific Characters.—Body of large size. Height of marginalia not twice their thickness. Median marginalia smooth. More distal marginalia with prominent mammilations.

Material.—There are about eight fragmentary specimens of this species in the British Museum (Nat. Hist.). The specimen registered E. 2576 (Mantell Coll.) shows considerable portions of the actinal surface, and that registered 35498 (Taylor Coll.) the dorsal view of a well-preserved portion of one arm. Both these are figured on Pl. XXIX. Another specimen, registered E. 2628 (Mantell Coll.), shows a portion of the abactinal surface of the disc. The other specimens are mostly collections of isolated plates. The specimen registered 35498 is taken as the type.

Description.—The abactinal surface of the disc appears to have been covered with a number of plates of generally uniform size, with an average diameter of about 3·8 mm. A few plates exceed this size, but in no case are they as large as the corresponding plates in *P. punctatus.*

No specimen is sufficiently well preserved to give the proportionate lengths of the major and minor radii, but there is no doubt that the arms were considerably produced. A row of hexagonal tabulate radialia are present throughout the greater portion of the arm. The breadth of the arm at the base in the specimen registered E. 2576 is about 22 mm. This specimen, however, judging by the dimensions of the marginal plates, does not by any means appear to have attained the usual size of the species. The length of its minor radius is 18·5 mm.

The median supero-marginalia are quite smooth and are distinguished from those of all other species of the genus by their thickness (Text-fig. 24 *a*). In full-grown specimens they appear to be 18 mm. in breadth, 6 mm. in length, and 10 mm. in thickness. More distally the supero-marginalia acquire large mammilate rugosities.

The infero-marginalia are similar in character to the superior series.

Two rows of ventro-lateralia enter the base of the arm. Most of the ventro-lateralia appear to have been rhomboidal in shape and of uniform size. They possess very shallow hexagonal fittings indicating the former possession of granules.

Locality and Stratigraphical Position.—Upper Chalk, Kent.

Genus—METOPASTER, *Sladen.* (See p. 30.)

9. Metopaster quadratus, n. sp. Text-figs. 1, 2, p. 98.

Specific Characters.—Marginal plates in interbrachial areas almost square. Raised area on marginal plates without spine-pits. Supero-marginal plates rugose on interior surface. Ultimate supero-marginal plates may or may not be the largest of the series, variation in this respect being especially marked. Abactinal plates of disc with distinct stellate marking.

Material.—There are three fairly perfect and four fragmentary specimens of this species in the collection of Dr. Blackmore, of Salisbury. Two of these are figured in Text-figs. 1, 2. The species was discovered by Dr. Blackmore, who suggested the specific name "*quadratus*" on account of the characteristic shape of the majority of the marginal plates. The type is the specimen figured in Text-fig. 2.

Description.—The abactinal area of the disc is covered with hexagonal plates, which have a distinctly stellate appearance on their upper surface.

The arms are distinctly produced. In the specimen figured in Text-fig. 1, $R : r : : 41$ mm. $: 28$ mm. In the specimen figured in Text-fig. 2, $R : r : : 41$ mm. :

TEXT-FIG. 1.   TEXT-FIG. 2.

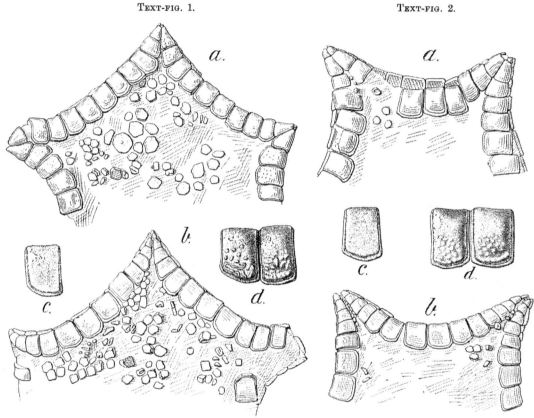

TEXT-FIG. 1.—*a*, Abactinal view of a specimen of *M. quadratus*, nat. size; *b*, actinal view of the same specimen; *c*, view of infero-marginal plate magnified two diameters; *d*, view of two supero-marginal plates magnified two diameters.

TEXT-FIG. 2.—*a*, Abactinal view of another specimen of *M. quadratus*, nat. size; *b*, actinal view of the same specimen; *c*, view of infero-marginal plate magnified two diameters; *d*, view of two supero-marginal plates magnified two diameters.

30 mm. The length of the side in the first-named specimen is 51 mm., in the second-named specimen 56 mm.

The supero-marginalia are either five or six in number, counting from the median inter-radial line to the extremity of the arm. In the inter-brachial area they are distinctly quadrate in character, and are from 7 to 8 mm. in width and from 6 to 7 mm. long.

The terminal paired supero-marginalia present very curious features. In the specimen figured in Text-fig. 1 some of these plates are large and triangular, just as in a typical *Metopaster*. Other terminal plates are, however, small and approximate to those characteristic generally of *Asteroidea*. The specimen figured in Text-fig. 2 presents no terminal plate which has a resemblance to those typical of *Metopaster*. In all other respects the specimens are almost exactly similar to one another.

The infero-marginalia are smooth and slightly concave in the centre. They are eight in number.

The actinal area of the disc is covered with sub-equal plates, which are four-sided in the inter-radial regions and tend to become hexagonal radially.

Locality and Stratigraphical Position.—Zone of *Actinocamax quadratus*, East Harnham, Salisbury.

Remarks.—The ornament of this species is identical with that of *M. uncatus*. The important differences between the species lie in the shape of the marginal plates, the character of the terminal supero-marginalia which show their especial peculiarities in all the specimens, and the ornament of the abactinal plates of the disc.

Family—PENTACEROTIDÆ (*Gray*), *emend. Perrier*, 1884. (See p. 76.)

Genus—STAURANDERASTER, novum. (See p. 125.)

12. S. ARGUS, n. sp. Pl. XXV, figs. 6, 6 *a* ; Pl. XXIX, figs. 8, 8 *a*, 9, 9 *a*.

Specific Characters.—Ossicles ocellato-punctate. Surface of ossicles very rarely truncate. If truncate, the flattened surface is not striated so as to simulate the asteroid madreporite.

Material.—Only very fragmentary specimens of this species are known. The best preserved specimen is in the collection of Dr. Blackmore, of Salisbury, and is figured on Pl. XXIX. Dr. Blackmore also possesses other specimens belonging to this species. Two specimens of the species are also preserved in the British Museum (Nat. Hist.), and bear the registered numbers E. 5019 and E. 2566 respectively. The first-named specimen was presented by Mr. W. McPherson. The second specimen was originally figured by Forbes in Dixon's 'Geology of Sussex,' pl. xxi, fig. 16, as a "fragment of an *Oreaster*." In Part III of this Monograph it was figured under my direction (Pl. XXV, figs. 6, 6 *a*) as Genus (?), Sp. (?). I have now isolated two or three ossicles from the specimen, and they are figured on Pl. XXIX, fig. 9. They show that the ossicles as originally figured merely present their interior aspect. The specimen E. 5019 is taken as the type.

15

Description.—The state of preservation of the fragmentary specimens of this species only allows adequate description of the ossicles of the disc. These are very uniform in character, and only differ from those of *Stauranderaster ocellatus* in the absence of the truncated summit with madreporiform striations. The largest ossicle measures about 6 mm. in diameter.

The isolated ossicles figured on Pl. XXIX show the characteristic shape of marginal ossicles belonging to the genus *Stauranderaster* (see p. 120), and assist us in ascribing not only this species, but also *S. ocellatus*, to which the species is closely allied, to that genus.

Locality and Stratigraphical Position.—The specimen presented to the British Museum (Nat. Hist.) by Mr. W. McPherson, is from the *Marsupites* zone at Brighton. The specimens in the collection of Dr. Blackmore are from Micheldever, Hants (zone of *Micraster cor-anguinum*).

Order—CRYPTOZONIA, *Sladen*, 1886.

Family—LINCKIIDÆ, *Perrier*, 1875.

Cryptozonate Asteroidea, with comparatively well-developed marginal plates, always contingent. Disc small, rays long and cylindrical. Abactinal skeleton tessellate. Tegumentary developments granulate, superambulacral plates usually present. Pedicellariæ (rarely present) excavate or foraminate.

Genus—LINCKIA, *Nardo*, 1834.

LINCKIA,   *Nardo*, 1834. De Asteriis, Oken's Isis, p. 717.
OPHIDIASTER (pars), *Müller and Troschel*, 1890. Monatsber k. preuss. Akad. Wiss. Berlin, p. 103.
LINCKIA,   *Gray*, 1841. Ann. Mag. Nat. Hist., vol. vi, p. 284.
ACALIA (sub. gen.), *Gray*, 1841. Tom. cit., p. 285.

Arms more or less cylindrical. Dorsal plates small, not arranged regularly in longitudinal series. Two or three rows of granules on the adambulacral plates. Superambulacral plates present. Papular areas distributed irregularly between the dorsal plates.

1. LINCKIA, ? sp. Pl. XXVII, figs. 1, 1 *a*.

Material.—A distorted specimen, which very probably belongs to the genus *Linckia*, is preserved in the British Museum (Nat. Hist.) (E. 5055, Capron Coll.).

Description.—The disc is small and very much distorted. The ventral aspect of one arm is the only portion of the star-fish which affords much opportunity for description. The arm is about 18 mm. long and 4 mm. broad, and possesses the cylindrical characteristic shape of the genus. It is composed of a large number of square ossicles, which are superposed in the cross-section of the arm, so far as it can be seen. They possess no spines, but a regular granulation appears to run lineally across their breadth.

Remarks.—This specimen is not sufficiently well preserved to ascribe it to a definite species of the genus.

Stratigraphical Position.—Lower Chalk.

CRETACEOUS OPHIUROIDEA.

Order—ZYGOPHIURÆ, *Bell* (1892).

Ophiuroidea, in which the movement of the ossicles on one another is limited by the development of lateral processes and pits. Superior, inferior, and lateral spine-bearing arm-plates are always present. The arms are simple and cannot coil round straight rods.

Family—OPHIOLEPIDIDÆ.

Zygophiuræ with oral papillæ from three to six, of which the last may be infradental. Arm incisures on the disc. Dental papillæ absent.

Genus—OPHIURA, *Lamarck*, 1801.

Ophiura,   *Lamarck.* Histoire Naturelle des Animaux sans Vertebres, vol. ii.
Ophiolepis,   *Müller and Troschel,* 1842. System der Asteriden (Braunschweig).
Ophioglypha, *Lyman,* 1865. (Ophiuridæ and Astrophytidæ) Ill. Cat. Mus. Comp.
                Zool. Harvard, Nos. i, viii; 1882, Rep. Challenger Zool.,
                vol. v.
        —   *Verrill,* 1899. Report on the Ophiuroidea collected by the Bahama
                Expedition in 1893; Iowa City, Bull. Labor. Nat. Hist., p. 4.

Disc covered with plates or scales which are for the most part swollen. Radial shields naked and swollen. Teeth. The inner mouth-papillæ long but

16

becoming smaller and shorter towards the distal oral region, where they are almost hidden by the scales of the mouth-tentacles. Arm-spines smooth and short, seldom longer than an arm-segment. Tentacle-scales numerous. The innermost pair of tentacle-pores narrow, surrounded by numerous tentacle-scales, and opening obliquely into the oral slits. In the back of the disc, where the arm joins it, a notch usually edged with papillæ. Two genital slits arise from the sides of the mouth-shields.

The following species are placed provisionally in this genus, to which the known characters would approximate them. The evidence, however, in every case is incomplete.

1. OPHIURA SERRATA, *Roemer*. Pl. XXVII, figs. 3, 3 *a*, 3 *b*, 3 *c*, 3 *d*, 3 *e*.

OPHIURA SERRATA, *Roemer, F. A.*, 1841. Die Versteinerungen norddeutsch. Kreidegeb., p. 28, pl. vi, fig. 23.

— — *Forbes*, 1843. Proc. Geol. Soc., vol. iv, p. 234, fig. 2.

— — *Reuss*, 1845–6. Die Verstein. böhm. Kreide-form., vol. ii, p. 58, pl. xx, fig. 26.

— — *Forbes*, 1850. In Dixon's Geology and Fossils of the Tertiary and Cretaceous Formations of Sussex, p. 337, pl. xxiii, figs. 2, 3, 3 *a*, 3 *b*.

— — *Morris*, 1854. Cat. Brit. Foss., ed. 2, p. 84.

— — *Forbes*, 1878. In Dixon's Geology of Sussex (new edition, Jones), p. 369, pl. xxiii, figs. 2, 3, 3 *a*, 3 *b*.

Specific Characters.—Conspicuous pear-shaped radials. Remainder of dorsal surface of disc covered with scales. Upper arm-plates occupying great proportionate width of arm. Six, occasionally seven ?, spines on each side of arm-segment.

Material.—Only two fragments of arms were originally available for description by Roemer. The specimens described by Forbes were more nearly complete, one specimen showing a considerable portion of the disc and the proximal portion of four arms (figured on Pl. XXVII of this Monograph), now in the British Museum (Nat. Hist.), and bearing the registered number E. 5043 (Dixon Coll.). There are fragments of arms in several collections which can apparently be ascribed to this species.

Description.—The disc is 15 mm. in diameter. Almost the whole of the dorsal covering has disappeared in the specimen figured, thus exposing the inner surface of the mouth-plates. The jaws (oral angle plates) are clearly seen. They are long and slender and do not meet inter-radially. The grooves for the water vascular canal and for the nerve-ring; and the depressions for the first mouth-tentacle and the entrance to the branch of the water vascular system are clearly shown. Fragments of the scaly covering of the disc are seen scattered over the disc. The peristomial plates are not recognisable, as is also the case in modern species of *Ophiura*. The arms are 3 mm. wide at the base. The upper arm-plates near the disc are broad. Six (or seven?) spines, which are in length about a third of the length of the arm-segment, are present.

The vertebral ossicles are figured on Pl. XXVII, figs. 3 *c*, 3 *d*, 3 *e*. They show the typical Ophiurid structure as displayed by modern species of Zygophiurids.

Locality and Stratigraphical Position.—Upper Chalk, Bromley, Kent.

2. OPHIURA FITCHII, n. sp., *ex Forbes*, MS. Pl. XXVII, figs. 2, 2 *a*, 2 *b*.

Specific Characters.—Body large and stoutly built. Disc covered with large swollen plates. Radials contiguous, large, kidney-shaped. Upper and lower arm-plates small.

Material.—An external cast in flint, the sole remains of this species, is preserved in the Norwich Museum (No. 2294).

Description.—Very little can be made out concerning the structure of this ophiurid. The disc appears to have been surrounded with a circlet of large radials. It has a diameter of about 16 mm.

A plasticine mould of the lower portion of the cast is figured on Pl. XXVII, fig. 2 *a*. The jaws (oral angle plates) are distinctly seen. They were long and slender, and similar in form to those of *O. serrata*. Similarly a peristomial plate is not visible. The inter-radial rounded buccal shield is clearly seen. The arm is 4·5 mm. broad at the base. The impression of the cast of the arm appears to indicate that both upper and lower arm-plates were small, as the side arm-plates appear to meet in the dorsal and ventral median lines.

Locality and Stratigraphical Position.—From flint gravel, Mousehold, Norwich.

3. OPHIURA PARVISENTUM, n. sp. Pl. XXVII, figs. 4, 4 *a*.

Specific Characters.—Disc covered with small plates, radials inconspicuous. Proximal upper arm-plates only occupying about one third of dorsal surface of arm. Five spines on each side of arm-segment.

Material.—There is only one specimen of this species. This is preserved in the British Museum (Nat. Hist.), and bears the registered number E. 5052 (purchased of Simmons).

Description.—The disc is about 15 mm. in diameter. It is covered with a large number of small plates, which are rather scattered. The arms are 3 mm. broad at the base. The upper arm-plate is much narrower than in *O. serrata.* There are five spines, about three quarters the length of an arm-segment, on each side of the arm.

Locality and Stratigraphical Position.—Upper Chalk, Bromley, Kent.

Genus—OPHIOTITANOS,[1] novum.

Disc covered with plates which are small and sub-equal. Radial shields small, triangular, naked, scarcely swollen. Arm-spines small. Mouth-shields large. Side mouth-shields small, widely separated.

1. OPHIOTITANOS TENUIS, n. sp. Pl. XXVIII, figs. 1, 1 *a*, 2, 2 *a*.

Proximal upper arm-plates longer than broad. Spines very short, five in number. Disc, with the exception of the radial plates, covered with an extensive granulation.

Material.—The material for the description of this species consists of several specimens in the British Museum (Nat. Hist.), The specimen E. 5056, which is the type, is figured on Pl. XXVIII, fig. 1, and the specimen E. 5057 on Pl. XXVIII, fig. 2. There are also specimens registered E. 5058, E. 5059, 57512. All are in a fair state of preservation. The first four specimens are from the Capron Coll., and the latter specimen was purchased from W. Griffiths. There are further examples of the species in the Sedgwick Museum, Cambridge.

Description.—The disc is flat on the dorsal surface, and its diameter in the largest specimen is 4·7 mm. Each pair of radial shields is separated by three ornamented plates.

The mouth-shields are almost oval in shape. The side mouth-shields are small, and lie widely separated on the outer edges of the mouth-shields. There

---

[1] Τίτανος = chalk.

is a large number of granules in the oral region, but the tips of the jaws (oral angle plates) can be seen just above the mouth-shields. There appear to be five (or six) squarish mouth-papillæ.

The first under arm-plate is small and with rounded edges. Distally the under arm-plates are at first almost square, then roughly pentagonal, and finally triangular, with the apex pointing towards the disc. The side arm-plates meet below at about the seventeenth arm-plate. The upper arm-plates are at first hexagonal, but rapidly become roughly triangular. They are tumid in appearance and have a rounded base. They rapidly become smaller, and allow the arm-plates to meet dorsally.

There are on each arm-segment five small smooth spines considerably shorter than the length of a segment.

There appear to be two tentacle-scales, but the exact number is rather difficult to determine.

Locality and Stratigraphical Position.—Lower Chalk, Folkestone and Dover.

2. OPHIOTITANOS LÆVIS, n. sp.   Pl. XXVIII, figs. 3, 3 *a*, 4, 4 *a*.

Specific Characters.—Spines longer than arm-segments. Upper arm-plates broader than long. Plates of dorsal surface of disc not hidden by granules.

Material.—Only one specimen is known. This is preserved in the British Museum (Nat. Hist.), no. E. 5053 (purchased of Mr. Griffiths), and shows the dorsal aspect.

Description.—The diameter of the disc is 4·8 mm., being thus about the same size as *Ophiotitanos tenuis.* The arm is, however, not so broad at the base, measuring here only 1·5 mm. across. The specific characters given above separate it sharply from this last-named species. I have not been able to determine the number of spines on each arm-segment.

Locality and Stratigraphical Position.—Lower Chalk, Dover.

Remarks.—A small specimen upon the slab, no. E. 5058 (Pl. XXVIII, figs. 4, 4 *a*), may be a young member of this species, as it possesses a strong general resemblance to the above. It is peculiar in having only four arms, probably an abnormality. As explained on p. 111, *Ophioglypha bridgerensis,* Meek, is very similar in appearance to this species.

3. Ophiotitanos magnus, n. sp. Pl. XXVIII, figs. 5, 5 *a*; Pl. XXIX, fig. 13.

Specific Characters.—Proximal upper arm-plates broader than long. Spines very short, seven spines on each arm-segment. Disc, with the exception of the radial plates, covered with an extensive granulation.

Material.—There are several specimens belonging to this species in the British Museum (Nat. Hist.) and in the Sedgwick Museum at Cambridge   The specimens in the British Museum bear the registered numbers E. 5060 (Capron Coll.), E. 5050, E. 370, and E. 371 (all from J. Starkie Gardner Coll.). The first-named speci-men, which shows the ventral aspect, is figured on Pl. XXVIII as the type. A specimen from the Sedgwick Museum showing the dorsal aspect is figured on Pl. XXIX.

Description.—This species is the largest of all the known Chalk Ophiuroids, the diameter of the disc being about 37 mm., and the breadth of the arm at the base 5 mm.   There appear to have been two tentacle-scales.

Remarks.—Portions of the arm of this species are very similar in form to those of *Ophiura serrata*. Unless the disc is present it is difficult to separate this species from that form.

Locality and Stratigraphical Position.—Lower Chalk.

Sub-order—NECTOPHIURÆ.

Spines situated at an angle to the arm.

Family—AMPHIURIDÆ, *Ljungman*, 1867.

Zygophiuræ with oral papillæ from one to five, of which the last is generally infradental. Arms inserted on ventral side of disc. Dental papillæ absent.

Genus—AMPHIURA, *Forbes*, 1842.

Disc small, delicate, covered with naked overlapping scales, and furnished with uncovered radial shields. Teeth. Mouth-angles small and narrow. Arms long, slender, even, and more or less flattened. Arm-spines short and regular.

1. AMPHIURA CRETACEA, n. sp.   Pl. XXVIII, figs. 6, 6 *a*.

Specific Characters.—Five (or six) mouth-papillæ. Two tentacle-scales. Five arm-spines.   Mouth-shields triangular in shape with a curved convex base, a fair proportion of the jaws (oral angle plates) showing on the ventral surface.

Material.—The one specimen known of this species is preserved in the British Museum (Nat. Hist.), no. E. 5059 (Capron Coll.).

Description.—The disc is 3 mm. in diameter.

There are five (or six) moderately stout blunt mouth-papillæ on either side of the mouth-angle, and a triangular papilla situated infradentally.   The side mouth-shields are long and narrow, with proximal and distal sides parallel.   They meet in the middle.   The extremities of the jaws are very obvious above these.

The arms are long and slender, and are 2 mm. broad at base.   The width of arm close to disc is 1·6 mm.   The first under arm-plate is small, with slight re-entering sides and a prominent median groove.   The remaining under arm-plates have a pointed proximal apex, and their sides are re-enteringly curved.   The distal side is convex.   The side arm-plates meet proximally on the under surface in the middle line.

The five stout tapering spines are rather longer than a joint of the arm, and are situated on a proximal ridge.

Locality and Stratigraphical Position.—Lower Chalk, Folkestone.

CRETACEOUS ASTEROIDEA AND OPHIUROIDEA FROM EXTRA-BRITISH LOCALITIES.

The following are the principal species from extra-British localities which have not been mentioned in the previous portions of the Monograph.   It will be seen that where accurate information is available, these species are almost entirely identical with British species.

ASTERIAS QUINQUELOBA, *Goldfuss*, 1826. 'Petrefacta Germaniæ,' p. 209, pl. lxiii, figs. 5 *a—u*.

The illustrations given of this species are very beautiful and accurate, and enable one to identify the ossicles ascribed to it as a mixture of ossicles of

Metopaster Parkinsoni (*a—p*), *Stauranderaster ocellatus* (*q—r*), and *Pentagonaster megaloplax* (*s—u*).

Various German writers have utilised the description of Goldfuss and this specific name for the identification of isolated ossicles—*e. g.* in Roemer, 1841, 'Die Verstein. norddeutsch. Kreidegeb.,' pl. vi, fig. 20, the ossicles ascribed to *A. quinqueloba* are really ossicles of *P. megaloplax*; while in Reuss, 1845—6, 'Verstein. böhm. Kreideform.,' p. 58, pl. xliii, figs. 15—20, and in Geinitz, 1872—5, 'Palæontographica,' vol. xx, pt. 2, pl. vi, fig. 7, the ossicles ascribed to *A. quinqueloba* are really ossicles of *M. Parkinsoni*.

Since the name *quinqueloba* is prior to all the other names mentioned above, it must be used instead of one of them. The simplest course appears to be to limit it in the sense of Roemer, by fixing on the specimen represented in Goldfuss's plate lxiii, fig. 5 *t*, *u*, as type of *Asterias quinqueloba*. The result of this is to replace the name *Pentagonaster megaloplax*, Sladen (antea p. 27, Pl. IV, figs. 2—4, Pl. XIII, figs.1 *a*, 1 *b*), by *Pentagonaster quinqueloba* (Goldfuss).

ASTERIAS JURENSIS, *Münster*, 1826. In Goldfuss, 'Petrefacta Germaniæ,' p. 210, pl. lxiii, figs. 6 *a—h*.

Figs. 6 *a—e* represent a fragment and isolated ossicles which closely resemble *Calliderma Smithiæ*, and figs. 6 *f—h* represent isolated ossicles bearing an equally strong resemblance to *Stauranderaster Boysii*. The species, however, is said to be " e calcareo jurassi Wurthembergia et Baruthino," whereas *C. Smithiæ* and *S. Boysii* are typical Cretaceous species, and have not in any other work been described from Jurassic rocks. Without seeing the original specimens no one should assert that Münster was so far mistaken as to the horizon and locality of the fossils before him. We can only suspend judgment.

ASTERIAS TABULATA, *Goldfuss*, 1826. 'Petrefacta Germaniæ,' p. 210, pl. lxiii. figs. 7 *a—g*.

Figs. 7 *a—b* are illustrations of isolated ossicles of *Stauranderaster argus* which are found in the Upper Chalk (zone of *Micraster. cor-anguinum*). I am unable to recognise the illustrations of the remaining ossicles as appertaining to any English Cretaceous species. The plates are said to be " e stratis argillaceis superioribus calcarei jurassi Baruthini." It is just possible that the locality and stratigraphical horizon are wrongly given in the case of *a* and *b*, and the name *Asterias tabulata* should be restricted to figs. *c—g*, one of those specimens being taken as type.

ASTERIAS SCHULZII, *Cotta*.

This is described and figured in Roemer, 1841, ' Die Verstein. Norddeutsch. Kreidegeb.,' p. 28, pl. vi, fig. 21, as follows : " Fünfeckig mit fünf kurzen Strahlen, unten vertieft und in der Mitte mit fünf Erhabenheiten ; der vorstehende Rand gewölbt und zwischen je zwei Strahlenspitzen mit etwa 45 schmalen Täfelchen besetzt."

The description reads as if the specimen belonged to the genus *Stauranderaster*, but neither this nor the figure given is much aid in the exact identification of the species.

A cast of a fossil Asteroid is ascribed to the same species under the name of *Stellaster Schulzii*, by Geinitz, ' Palæontographica,' vol. xx, pt. 2, pl. v, figs. 3, 4. This cast, however, looks like a cast of a species of either *Calliderma* or *Nymphaster*.

ASTERIAS? DUNKERI, *Roemer*, 1841. ' Die Verstein. norddeutsch. Kreidegeb.,' p. 27.

This species is described as follows : " Die Flächentäfelchen sind länger als breit, 4—6 eckig, schrägrandig fein gekörnt und nahe am obern Rande durchbohrt." Only isolated plates were known. I am unable to identify the species with other known Asteroids. The illustration given is not very clear.

According to Roemer, these isolated plates were described by Dunker and Koch, 1837, ' Norddeutsch. Oolithgeb.,' as plates of *Cidaris variabilis*.

CŒLASTER COULONI, *Agassiz*, 1836. ' Mém. Soc. Sci. Nat. Neuchatel,' vol. i, p. 191.

No figure was given of this species by Agassiz. A fossil Asteroid figured by Fritel, ' Le Naturaliste,' vol. xvi, 1902, p. 79, as this species, appears to be somewhat like *Nymphaster marginatus*.

CUPULASTER PAUPER, *Frič*, 1893. ' Arch. Landesdf. Böhmen,' ix, no. 1, p. 112.

This species is named from a specimen 3 mm. in diameter, which so obviously presents the large terminal plates which are common to all very young forms of starfish, that it is useless to speculate as to its identity.

Locality and Stratigraphical Position. — Cretaceous (Priesener Schichten), Waldek, near Bensen, Bohemia.

17

GONIASTER MARGINATUS, *Reuss*, 1845—6. 'Verstein. böhm. Kreideform.,' p. 58,
pl. 43, figs. 21—32.

The illustrations of the isolated ossicles described as this species bear strong
resemblance to those of *M. Parkinsoni*.

GONIASTER MAMMILLATA, *Gabb*, described by *Clark*, 1892, in 'Bull. U.S. Geological
Survey,' no. 97, p. 32.

" *Determinative Characters.*—Body pentagonal, provided with a dorsal and a
ventral row of marginal plates that are narrower than high, and distinctly tumid
on their outer surface.   Only detached marginal plates preserved."

Remarks.—The isolated plates undoubtedly belong to a species of *Pycinaster*
and bear a strong resemblance to those of *Pycinaster angustatus*.

" *Locality and Geological Position.*—Yellow Limestone of the middle marl bed
of the Cretaceous from Vincentown, New Jersey."

PENTACEROS DILATATUS, *S. Meunier*, 1906.   'Le Naturaliste,' (2), vol. xx, p. 117.

The specimen described under this name is an external cast in flint, of which
a plaster cast has been presented to the British Museum (Nat. Hist.), E. 13075.
Owing to the courtesy of Professor Stanislas Meunier I have been able to examine
the original specimen.   It shows a well-preserved impression of the abactinal
surface.   The ossicles of the disc are rhomboidal or hexagonal, contiguous, of
almost uniform size, about 2 mm. in diameter.   The specimen, therefore, cannot
be placed in the genus *Pentaceros*, and its appearance, measurements, and type
of ornament enable me to ascribe it to *Pentagonaster obtusus*.

OPHIOGLYPHA BRIDGERENSIS, *Meek*, described and figured by *Clark*, 1892.   'Bull.
U.S. Geological Survey,' no. 97, p. 29.

' *Determinative Characters.*—Disc composed of numerous small imbricating
plates.   Upper arm-plates wider than long, the outer angles sharp and expanding
between the side arm-plates, which are slightly smaller.   Under arm-plates long
and nearly rectangular in shape.

"*Dimensions.*—Diameter of disc 6 mm. Length of arm 20 mm. Width of arm near disc 1·25 mm."

Remarks.—The figures given and dimensions of this species appear to me to indicate that it closely resembles *Ophiotitanos lævis.* The upper arm-plates are described as hexagonal, and the more proximal upper arm-plates of the British species present this appearance.

Locality and Stratigraphical Position.—Cretaceous, Fort Ellis, Montana.

OPHIOGLYPHA TEXANA, *Clark*, 1892. 'Bull. U.S. Geol. Survey,' no. 97, p. 30.

"*Determinative Characters.*—Disc round; composition indistinct. Arms long, with wedge-shaped under arm-plates about as wide as long; upper arm-plates about twice as wide as long.

"*Dimensions.*—Diameter of disc 15 mm. Length of arm 50 mm. Width of arm at disc 2 mm."

Remarks.—It is difficult to identify this with a British species, but the illustrations appear to indicate that it is somewhat similar to *Amphiura cretacea.*

Locality and Stratigraphical Position.—Cretaceous, six miles north of Fort Worth.

OPHIURA GRANULOSA, *Roemer*, 1841. 'Die Verstein. norddeutsch. Kreidegeb.,' p. 28, pl. vi, fig. 22.

This species is described as follows:
" Die Arme sind walsenförmig und bestehen aus gewölbten, seitlich durch ein Furche getrennten, deutlich gekörnten Seitenschildchen; wo sich deren vier berühren liegt ein kleines, dreieckiges Schildchen dazwischen."

Remarks.—The small fragment of the arm figured shows it to be a portion of the distal extremity—quite possibly the distal extremity of some species of *Ophiotitanos.*

Locality and Stratigraphical Position.—Lower Chalk, Lindener Berg, near Hanover.

Ophiura ? pulcherrima, *Fric*, 1893.　'Arch. Landesdf. Böhmen,' vol. ix, no. 1, p. 113.

Remarks.—No description is given.　The figure does not show any resemblance of this species to. other known forms, but the specimen was obviously imperfect. The upper arm-plates appear to be **V**-shaped.

Locality and Stratigraphical Position.—Cretaceous (Priesener Schichten), Waldek, near Bensen, Bohemia.

Stellaster albensis, *Geinitz*, 1872—5.　'Palæontographica,' vol. xx, pt. 2, p. 16, pl. vi, fig. 3.

Remarks.—This species is only known from a cast from the Quadersandstein. In the absence of determinative characters it is impossible to say whether it is identical with or differs from more fully described species.

Stellaster Coombii, *Forbes*, sp., in *Geinitz*, 1872—5, 'Palæontographica,' vol. xx, pt. 2, p. 17, pl. vi, figs. 4—6.

The specimens illustrated here as *S. Coombii* certainly do not belong to Forbes' species of that name.　They appear to be ossicles of various species, but I am unable to identify them from the figures given.

SPECIFIC AND GENERIC CHARACTERS IN CHALK ASTEROIDEA.

When I commenced this account of Cretaceous Asteroids I endeavoured, so far as possible, to follow the generic classification of previous authors and especially to preserve the continuity of Mr. Sladen's work.　More recent work, however, has led me to believe that the shape of the marginal plates, together with their ornament, affords the best determinative generic and specific characters, and further enables us to identify almost all Cretaceous starfishes from single isolated plates.

Some necessary revision as to nomenclature in both genera and species is given below, together with an illustrated key-table, which it is hoped will enable zonal collectors to identify the isolated asteroid plates which are commonly met with in almost all exposures. Up to the present our knowledge of the zonal occurrence of these forms has been limited, as complete specimens of starfishes are exceedingly rare. There appears to be no reason now, however, why our knowledge of the zonal distribution of these forms should not become as nearly complete as it is, for example, in the case of Echinoids.

I must thank Dr. Blackmore, of Salisbury, for his invaluable suggestions to me concerning this means of identification.

Ornament.—The ornament of starfishes consists of calcareous pieces, which may be spinous in form, or scaly, or granular. These may occur:

(1) Embedded in the living tissues outside the general body-plates, but not in contact with the plates themselves. On the death of the animal they become dispersed on the disintegration of the living tissues, and such ornament is therefore rarely visible in fossil specimens.

(2) Articulated to the plates. In this case they are situated:

Text-fig. 3.—Isolated marginal of *Nymphaster Coombii*, showing spines and spine-pits.

Text-fig. 4.—Isolated marginal of *Stauranderaster bulbiferus*, showing the "pustulate" type of ornament.

(*a*) Either in a depression of the plate;

(*b*) Or in a depression upon a raised eminence of the plate. Occasionally in this latter case the depression may be excavate in the centre in order to allow a strong muscular attachment. In this case the eminence may simulate the perforate tubercle of an Echinoid such as *Cidaris*.

In almost all cases in Cretaceous Asteroids the ornament is of the type 2 *a*.

Generally the movable articulated pieces have disappeared, but in such cases the depression on the plate which they formerly occupied is readily visible (compare Text-fig. 3).

I purpose to call all such movable articulated pieces, whether they are spinous or granular in character, "spines," and, at the suggestion of Dr. Bather, the depressions on the plate "spine-pits."

The character of the spine-pits appears to be constant in character in each individual species. Thus, *e. g.* in *N. Coombii* (Text-fig. 3) they show a coarse honeycomb structure, uniform in character over the whole of the plate. In *Stauranderaster bulbiferus* (Text-fig. 4) the spine-pits are circular and widely spaced. This latter type is interesting, as it apparently occurs only in the genera *Metopaster* and *Stauranderaster*. The spine is very small, and barely projects over the edge of the deep spine-pit, thus giving the plates an embossed appearance. This type of ornament I call the "pustulate" type.

In previous portions of this Monograph both Mr. Sladen and myself have assumed that if no spine-pits are present on a plate they have been weathered

TEXT-FIG. 5.—Pedicellaria of *Nymphaster oligoplax.*

TEXT-FIG. 6.—Three pedicellaria from *Pentagonaster quinqueloba* on the right. A pedicellaria from *Hadranderaster abbreviatus* on the left.

TEXT-FIG. 7.—A pedicellaria from *Metopaster Parkinsoni* in the centre, on the left a pedicellaria of *Pycinaster senonensis*, on the right a pedicellaria of *Stauranderaster coronatus.*

away. It now appears that the absence of spine-pits is such a constant character in certain species that this supposition can no longer be held, and the absence of spine-pits indicates an original absence of spines, or, at any rate, spines articulated to the plates. In support of such a conclusion it can be urged that, generally speaking, Chalk fossils are but little weathered, and that there is evidence derived from a study of recent forms.

Pedicellariæ.—As can readily be understood, only pedicellariæ which are articulated in depressions of the plate are preserved in Cretaceous Asteroids. Purse-like, valvate pedicellariæ (Text-fig. 5) of a generalised type are common to many genera. More specialised pedicellariæ, however, peculiar to the genera *Metopaster* and *Pycinaster* are also met with (Text-fig. 7).

KEY-TABLE FOR THE IDENTIFICATION OF CRETACEOUS ASTEROIDS.

The following key-table is based on the shape of the marginal plates and the character of the spine-pits on them. Generally speaking both superior and inferior series are similar in these respects, but when otherwise a note is made in the table.

A short description is also given of various plates which cannot be adequately treated in the table.

All the plates figured in the table are magnified 4 diameters.

It is convenient to consider the Chalk (Cenomanian-Senonian) species separately from the Upper Greensand forms. No Cretaceous Asteroidea have been described from below this horizon.

CHALK (CENOMANIAN-SENONIAN) SPECIES.

I.—Marginals four-sided, with sides rectilinear or almost rectilinear; broad.

    A. Without a rabbet-edge. *Calliderma, Nymphaster, Pentagonaster.*
    B. With a rabbet-edge. *Metopaster, Mitraster.*
    C. With a distinct ridge. *Arthraster.*

II.—Marginals either hexagonal or rounded; very thick. *Hadranderaster.*

III.—Marginals wedge-shaped, high, spine-pits very shallow or absent. *Pycinaster.*

IV.—Marginals breast-plate-shaped. *Stauranderaster.*

V.—*Miscellaneous plates.*

I.—Marginals four-sided, with sides rectilinear or almost rectilinear. Pedicellariæ when present of a simple bi-valvate character.

    A. Without a rabbet-edge. *Calliderma, Nymphaster, Pentagonaster.*

        1. Spine-pits shallow, hexagonal, giving a honeycomb appearance.

(A)   (B)

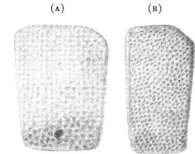

TEXT-FIG. 8.

a. Honeycomb medium or fine. *C. Smithiæ* (see p. 123).
    Text-fig. 8 A.—Variety with medium-sized spine-pits (see p. 123).
    Text-fig. 8 B.—Variety with fine spine-pits (see p. 123).

b. Honeycomb coarse. *N. Coombii.* Text-fig. 3, p. 113.

c. Honeycomb confined to a portion of the plate (or absent). *P. obtusus.* Text-fig. 9.

TEXT-FIG. 9.

2. Spine-pits shallow, circular, circles variable in size, adjoining.

One species, *C. latum.*   Text-fig. 10.

TEXT-FIG. 10.

3. Spine-pits deep, circular, not adjoining.

a. Spine-pits coarse.   *N. marginatus.*   Text-fig. 11.
b. Spine-pits fine, not on margin.   *N. oligoplax.*   Text-fig. 12.

TEXT-FIG. 11.   TEXT-FIG. 12.

c. Spine-pits fine, uniformly over the whole of plate.   *P. lunatus.*
Text-fig. 13.

TEXT-FIG. 13.

4. Spine-pits on outer edge of plate with raised margins.

One species, *P. quinqueloba* (see p. 108).   Text-fig. 14.

TEXT-FIG. 14.

5. No spine-pits present.

(A) (B)

Text-fig. 15.

a. Proximal marginalia almost smooth; distal with granular rugosities arranged in a linear series. *N. radiatus.* Text-fig. 15, (A) proximal marginal, (B) distal marginal.

Text-fig. 16.

b. Marginalia with rugosities not arranged in linear series. *N. rugosus.* Text-fig. 16.

B. Marginals with a rabbet-edge. Rabbet-edge covered with small spine-pits. Pedicellariæ when present "winged" (see Text-fig. 7).

1. Spine-pits on central raised area.

(A)

(B)

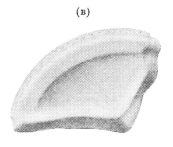

Text-fig. 17.

a. Central raised area smooth on both superior and inferior marginalia. Spine-pits on infero-marginalia uniformly situated. *M. Parkinsoni.* Text-fig. 17, (A) outer view of marginal, (B) side view of marginal.

(B)

(A)

Text-fig. 18.

b. Central raised area of supero-marginalia rugose; that of infero-marginalia smooth, with spine-pits in form of network. *M. Hunteri.* Text-fig. 18, (A) supero-marginal, (B) infero-marginal.

c. Central raised area of supero-marginalia smooth in young, rugose in mature individuals. Lower Chalk form. *M. cornutus* (see p. 124).

18

2. No spine-pits on central raised area.

(A)   (B)

a. Rugosities present on greater portion of surface of supero-marginalia. *M. rugatus*. Text-fig. 19, (A) supero-marginal, (B) infero-marginal.

TEXT-FIG. 19.

(A)   (B)

b. Rugosities confined to inner edge of supero-marginalia or absent; outer portion of plate tumid. *a*. Plates oblong. *M. uncatus*. Text-fig. 20, (A) supero-marginal, (B) infero-marginal. *β*. Plates square. *M. quadratus*.

c. No rugosities present on the supero-marginalia which are not tumid on their outer portion. *M. compactus*. See Pl. XXVI, fig. 3 *b*.

TEXT-FIG. 20.

N.B.—The infero-marginalia of the above species are difficult to distinguish except by their dimensions. The reader is advised to refer to the detailed description for these.

c. Marginalia with a distinct ridge which has granular elevations along its base. *Arthraster*.

(A)   (B)

1. Upper surface of ridge smooth. *A. Dixoni*. Text-fig. 21, (A) outer view of marginal, (B) side view of marginal.

2. Upper surface of ridge with spine-pits. *A. cristatus*. See Pl. XXIX, fig. 10.

TEXT-FIG. 21.

II.—**Plates either hexagonal or rounded ; very thick. Spine-pits form a distinct, fine honeycomb marking.** *Hadranderaster* (see p. 125).

(A)   (B)

There is only one species. *H. abbreviatus*. Text-fig. 22, (A) outer view of marginal, (B) side view of marginal.

TEXT-FIG. 22.

III.—**Marginals generally wedge-shaped, high, spine-pits very shallow or absent.** **Pedicellariæ when present with five valves round a deep central depression.** *Pycinaster.*

(A)   (B)   (c)

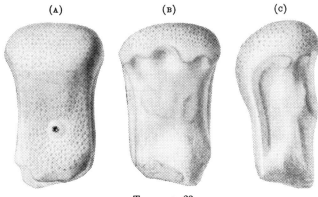

TEXT-FIG. 23.

1. Plates of maximum height, 10 mm. *P. angustatus.* See Pl. IX, fig. 1 *a*.
2. Plates of maximum height, 20 mm., ↶-shaped in profile. *P. senonensis.* Text-fig. 23, (A) outer view of marginal, (B) interior view of marginal, (c) side view.

TEXT-FIG. 24.

3. Plates very thick, often oblong, not wedge-shaped. *P. crassus.* Pl. XXIX, fig. 3, exterior view of marginal. Text-fig. 24, side view of marginal.

IV.—**Marginal plates breast-plate-shaped.** *Stauranderaster* (see p. 125).

A. With spine-pits.

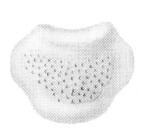 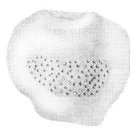

TEXT-FIG. 25.   TEXT-FIG. 26.

1. Spine-pits fairly coarse. *S. bulbiferus* (commonly occurring). Text-fig. 25.
2. Spine-pits medium. *S. bipunctatus* (only one specimen known). Text-fig. 26.

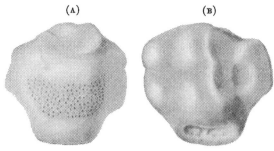

(A)   (B)

TEXT-FIG. 27.

3. Spine-pits fine. *S. Boysi.* Text-fig. 27, (A) outer view of marginal, (B) interior view of marginal.

B. Without spine-pits.

TEXT-FIG. 28.

TEXT-FIG. 29.

1. Maximum size about 7 mm. *S. coronatus.* Text-fig. 28.
2. Maximum size about 2·5 mm. *S. squamatus.*
3. Maximum size about 2·5 mm. *S. pistilliferus.* Text-fig. 29.

N.B.—It is impossible to distinguish isolated marginalia of these two species. Compare V, 3 (p. 121).

V.—*The collector may also come into possession of the following plates:*

1. Large triangular plates having the characteristic ornament of *Metopaster.* These are the ultimate supero-marginalia which characterise the genus. See *e. g.* Pl. XVI, fig. 2 *a*.

2. Large hemispherical plates with a flattened base. These are met with in the following species:

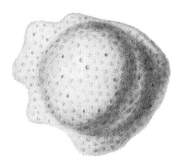

TEXT-FIG. 30.

A. With spine-pits.

Stauranderaster bulbiferus, S. Boysi, and ? *S. bipunctatus.* These are distinguished from one another by their spine-pits, which are of the same character as those met with in the marginalia. Text-fig. 30= a primary inter-radial of *S. bulbiferus.*

B. Without spine-pits.

Pycinaster senonensis, distinguished by the smooth surface, Pl. XXIX, fig. 6.

Arthraster Dixoni, distinguished by the rugose surface, Pl. XXIX, fig. 11.

3. Nodular plates with an excavate margin met with in the following species :

   A. With flattened base, from 7—9 mm. in diameter, *S. coronatus*, Pl. XXIV,
                                                            fig. 2.

        ,,   ,,   ,, 3—4 mm.   ,,   *S. squamatus*, Pl. XXV,
                                                              fig. 3.

   B. With produced base, *S. pistilliferus*, Pl. XXV, fig. 5.

4. Oblong or almost oblong plates having the characteristic ornament of *S. bulbiferus*. These are the more distal marginals, or in some cases the ventro-lateralia of this species. Generally the marginals are indented at the corners, and their shape can be decided from the characteristic breast-plate form.

                             Text-fig. 4, p. 113.—Outer view of plate.
                             Text-fig. 31.—Side view of plate.

     Text-fig. 31.

5. Irregularly rounded or polygonal plates with ocellate depressions having a raised ridge.

                 A. Without madreporiform markings on summit. *S. argus.* Pl. XXIX,
                     fig. 8 a.
                 B. With madreporiform markings on summit. *S. ocellatus.* Text-
                     fig. 32.

     Text-fig. 32.

6. One specimen of *Astropecten?* sp. and *Linckia* sp.? respectively have also been described, see pp. 90 and 100.

Upper Greensand Forms.

The genus *Comptonia* appears to be characteristic of this horizon. The plates are very similar in shape to those of *Calliderma*, except that they are more rounded in profile. Unfortunately, no specimen at present known shows the ornament of the marginalia.

I have also met with *C. Smithiæ* and *P. punctatus* from this horizon. They are readily distinguished by the characters which have already been given in the key-table.

NOTES ON THE KEY-TABLE.

Genera—CALLIDERMA, NYMPHASTER, PENTAGONASTER.

The following are the chief distinctive characters which separate these three genera in recent forms.

CALLIDERMA.	PENTAGONASTER.	NYMPHASTER.
1. Arms well produced.	Arms slightly produced.	As in *Calliderma*.
2. Abactinal area covered with closely fitting plates.	As in *Calliderma*.	Abactinal area covered with paxillæ, which are not closely fitting in the radial areas.
3. Ventro-lateral and infero-marginal plates with prominent spines.	Ventro-lateral and infero-marginal plates without prominent spines.	As in *Pentagonaster*.
4. Armature of the adambulacral plates consisting of 14–16 small spines arranged uniserially, with three or four rows of larger spines arranged rather irregularly.	Armature of the adambulacral plates arranged in longitudinal series. Series on the whole uniform in character.	As in *Pentagonaster*.

The fossil species of *Calliderma* possess the characters numbered 1 and 2, but differ to a greater or less extent in characters 3 and 4, in which they resemble *Pentagonaster*. The genus *Calliderma* was, however, founded by Gray on one species—*C. emma*. No other recent species has been assigned to the genus. It is difficult, therefore, to say how far the distinctive characters of the Cretaceous genera should have generic value. The question is debated by Mr. Sladen on p. 5 of this Monograph, and the very striking general resemblance of the fossil species to the recent *C. emma* influenced him in his decision to group them under this genus. There appears no great reason to dispute this assignment, but I am in more doubt as to the systematic position of the species which have been placed in the genera *Pentagonaster* and *Nymphaster*. It will be seen that as the fossil species of *Calliderma* resemble *Pentagonaster* in characters 3 and 4, the only distinctive character which remains between the two genera is the length of the arm. Two of the fossil species of *Pentagonaster*, namely *P. lunatus* and *obtusus*, have all arms which are well produced (the arms in the specimen of *P. lunatus* figured on Pl. IV, fig. 1, are broken off short). The third and remaining species, *P. quinqueloba*, is usually much more pentagonal in shape, although a specimen in the possession of Dr. Blackmore has a major radius at least twice the magnitude of that of the minor radius.

The species assigned to the genus *Nymphaster* by Sladen were so assigned because their structure and character, so far as they could be made out from the fragmentary condition of the fossils, appeared to warrant their inclusion in the genus *Nymphaster* (see p. 15).

It appears to me that these species have the same generalised characters as those assigned to the genera *Calliderma* and *Pentagonaster*. The distinctive character of the genus *Nymphaster* is the possession of paxillæ on the abactinal plates. No fossil species is sufficiently well preserved to show whether these were absent or present, and it is impossible therefore to confirm or deny Sladen's suggestion.

It will be seen from the above that there is no certain evidence which entitles us to distribute the Cretaceous species amongst the three genera, and it may be the task of a future observer to place them in one new genus. I have, however, in order to secure uniformity, utilised all these generic names even for the description of new species. The following species also appear to require revision.

Calliderma Smithiæ, C. mosaicum.

After examination of the fairly numerous specimens of the fossils assigned to these species in the British Museum (Nat. Hist.) I cannot confirm the specific distinctions made by Sladen on pp. 10 and 11 of this Monograph. All the characters mentioned vary greatly in individual specimens. The ornament, however, is common in character to both species, and I should prefer to unite them in one species, namely, *C. Smithiæ*, as this has prior place in the original account given by Forbes.

The specimens figured on Pl. VII, figs. 1 and 2, and stated by Sladen to be in his opinion doubtful examples of *N. Coombii* (p. 17), should in my opinion be assigned to *C. Smithiæ*, as should also the specimen figured on Pl. XIX, fig. 3. These examples possess a finer type of honeycomb structures on their marginalia than is usually met with in *C. Smithiæ*, and they may, therefore, be a distinct variety of this species (see Text-fig. 8).

A specimen preserved in the British Museum (Nat. Hist.) E. 5063, was figured by Sladen on Pl. V, fig. 1 *a*, of this Monograph as *Tomidaster sulcatus*. Apparently it was the intention of Sladen to make a new genus and species for the reception of this fossil. The dimensions of the marginalia, compared with the minor radius and the ornament, are precisely the same as in *C. Smithiæ*, from which the specimen only differs in the possession of numerous valvate pedicellariæ. In view of the somewhat freakish way in which pedicellariæ occur in starfishes, it does not appear to me that this character alone entitles us to make a specific or generic distinction on behalf of this specimen.

There is another specimen also in the collection of the British Museum (E. 1116) which has similarly numerous valvate pedicellariæ, but which shows the abactinal aspect. In all other respects the specimen cannot be distinguished from a specimen of *C. Smithiæ*.

Genus—METOPASTER.

In the key-table I have only distinguished four species of *Metopaster*, namely, *M. Parkinsoni*, *M. uncatus*, *M. quadratus*, and *M. cornutus*, the latter being a doubtful species. If one examines collections of Cretaceous Asteroids, one finds that practically all the specimens have been rightly assigned to these species. Specimens which could be assigned to the species *M. Mantelli*, *M. Bowerbanki*, *M. zonatus*, *M. sublunatus*, *M. cingulatus* (see pp. 38–55), are very rarely met with.

TEXT-FIG. 33.—Marginal of *Metopaster Parkinsoni*, showing a more scattered type of ornament than that usually met with.

The very considerable variation which occurs in undoubted specimens of *M. Parkinsoni* in the number of the supero-marginalia, their form, amount of ornament, and the shape of the ultimate plates of this series, makes specific characters founded solely upon these characters of doubtful validity, particularly as such variations occur even in an individual specimen, and it is upon a rather extreme variation of these characters occurring in very few specimens that this large number of species have been described. On the other hand, the presence or absence of spine-pits on the raised central area of the plate is a constant character in species of *Metopaster*. Two specimens figured in the Monograph appear to belie this statement. The specimen figured on Pl. X, fig. 4 a, shows no spine-pits on its infero-marginalia, but is figured as *M. Parkinsoni*. I have isolated a dorsal ossicle, which shows the specimen undoubtedly to belong to *M. uncatus*. The specimen figured on Pl. XI, fig. 3 a, as *M. uncatus*, shows spine-pits on its supero-marginalia in one inter-radius only. After very careful examination of this specimen, I have come to the conclusion that this inter-radius —the right-hand upper inter-radius of the figure—has been added by a dealer from a collection of ossicles of *M. Parkinsoni* to an imperfect specimen of *M. uncatus*.

Genus—Stauranderaster,[1] novum.

Pentaceros (pars). Pp. 76–89 of this Monograph.

The species *bulbiferus, Boysii, coronatus, ocellatus, bispinosus, pistilliferus,* and *squamatus* (pp. 76—89 of this Monograph), which have been formerly placed in the genus *Pentaceros,* together with the new species *argus* (p. 99), should, I think, now be ascribed to a new genus. The plates of these species are breast-plate-shaped, at times almost cross-shaped, and bear a characteristic type of ornament (see p. 113). In both these respects and in the absence of papular areas the species differ widely from species of recent *Pentaceros,* with which the only feature they have in common is the circlet of raised plates on the abactinal surface of the disc.

The type species of the new genus is *Stauranderaster Boysii,* and its diagnostic characters are :

Arms high, well produced, marginalia breast-plate-shaped, generally with a rabbet-edge free from ornament, and a raised interior area, either smooth or with ornament of an embossed type. A circlet of swollen plates present on the abactinal surface of the disc.

This genus, with the following genus, may be placed provisionally in the family Pentacerotidæ.

Genus—Hadranderaster,[2] novum.

Pentaceros (pars), p. 86 of this Monograph.

The species described as *Pentaceros abbreviatus* on p. 86 of this Monograph differs so considerably in the shape of its marginalia and their ornament from the species of *Stauranderaster* and recent species of *Pentaceros* that I have placed it in a new genus.

The type species is *Hadranderaster abbreviatus,* and the diagnostic characters of the new genus are :

Arms high, well produced, marginalia either hexagonal or rounded, very thick, ornament spread uniformly over the surface of the plate. Pedicellaria bi-valvate.

---

[1] σταυρός = a cross, ἄνδηρον = a raised garden border.
[2] ἁδρόι = stout, ἄνδηρον = a raised garden border.

GENERAL CHARACTERISTICS OF CRETACEOUS ASTEROIDEA AND OPHIUROIDEA.

The majority of Cretaceous starfishes belong to the Phanerozonate forms included in the families Pentagonasteridæ and Pentacerotidæ. Modern forms of the genera of these families are widely distributed geographically, but, generally speaking, they are characteristic of warmer waters than those of the English Channel of to-day.

The Chalk starfishes are specialised types which, although approximating to, are not identical with, modern genera. The differences, at any rate in some cases, appear to be distinctly physiologically advantageous.

Metopaster and Mitraster, the most abundant of all Chalk starfishes, possess not only a specialised type of ornament but also characteristic massive plates. The arms tend to become shortened and the disc correspondingly enlarged.

The Chalk species of the Pentacerotidæ differ from the modern forms, inasmuch as they are more strongly built; the abactinal areas are not reticulate, and all species possess intermarginalia which cause the characteristically deep body of these forms.

The Chalk is a deposit formed in seas which were sufficiently distant from land to avoid any great admixture of clay or sand. *Globigerina* and other forms of pelagic Foraminifera floated in abundance on the surface of the sea, which, because of its temperature, must have been exceedingly favourable to prolific forms. In the circumstances there must have been an abundance of food for starfishes, and we find, therefore, that the long-armed, comparatively active Astropectinidæ, which were so characteristic of the Jurassic shallow water deposits, are displaced by more sedentary forms which tend to specialise, so as to obtain, by the enlargement of the disc or development of intermarginalia, the largest possible space for their digestive organs.

The irregular Echinoids which are so characteristic of the Cretaceous seas are similarly sedentary forms.

The fossil Ophiuroidea also closely resemble modern forms. The isolated vertebral ossicles of *O. serrata,* figured Pl. XXVII, figs. 3 c, 3 d, 3 e, cannot be distinguished from the ossicles of recent Ophiuroids. Complete specimens of Ophiuroidea and Asteroidea are rare, but isolated plates are very numerous in the Upper Chalk. They are, on the contrary, rare in the Lower Chalk, according to experienced collectors, as, for example, Mr. Dibley.

The following starfishes are found in the zones indicated. The list is compiled from the papers of Dr. Rowe ('Proc. Geol. Assoc.,' vols. xvi, xvii), and also from notes furnished by Dr. H. P. Blackmore and Mr. T. H. Withers.

	Zone of Ostrea-lunata.	Zone of Belemnitella mucronata.	Zone of Actinocamax quadratus.	Zone of Marsupites.	Zone of Micraster cor-anginum.	Zone of M. cortes-tudinarium.	Zone of Holaster planus.	Zone of Tere-bratula gracilis.	Zone of Rhynchonella Cuvieri.	Zone of Holaster subglobosus.	Zone of Ammo-nites varians.
Nymphaster Coombei, Forbes	...	...	...	...	...	...	...	...	×	...	...
— oligoplax, Sladen	...	...	...	...	×	...	...	...	...	...	...
Pentagonaster quinqueloba, Goldfuss	...	×	×	×	×	...	...	...	...	...	...
— obtusus, Forbes	...	...	×	×	...	...	...	...	...	...	...
Metopaster Parkinsoni, Forbes	...	...	×	...	×	×	×	×	...	...	...
— uncatus, Forbes	...	...	...	×	×	×	...	...	...	...	...
— quadratus, n. sp.	...	...	×	...	...	...	...	...	...	...	...
— cornutus, Sladen	...	...	...	...	...	...	...	...	×	...	...
Mitraster Hunteri, Forbes	...	×	...	...	×	...	...	...	...	...	...
— rugatus, Forbes	...	×	×	...	×	...	...	...	...	...	...
Pycinaster angustatus, Forbes	...	×	×	×	×	...	...	...	...	...	...
— senonensis, Valette	...	...	×	...	...	...	...	...	...	...	...
Stauranderaster bulbiferus, Forbes	...	×	×	×	×	×	...	...	...	...	...
— Boysi, Forbes	...	...	×	...	...	...	...	...	...	...	...
— ocellatus, Forbes	...	×	×	×	×	...	...	...	...	...	...
— pistilliferus, Forbes	...	×	×	...	...	...	×	...	...	...	...
— argus, n. sp.	...	...	...	...	×	...	...	...	...	...	...
Arthraster Dixoni, Forbes	...	...	...	×	...	...	...	×	...	...	...
,, cristatus, n. sp.	...	...	...	...	×	...	...	...	...	...	...

THE PHYLOGENY OF THE CRETACEOUS ASTEROIDEA.

If we examine the various species of a genus or group of related genera of Cretaceous Asteroids, we find that there is a similar transition from smooth to spinous forms through an intermediate form, to that which has been observed in Ammonites and Brachiopods.

Both in Ammonites and Brachiopods single specimens show the whole life-history of the individual, for the shell of the animal is not materially altered in character after it has once been formed. It is therefore possible to show, e. g., that the character of the ornament of the shell of an Ammonite was smooth in its infancy, costate in its adolescence, spinous in the adult, and it has also been shown that this life-history depicted by the individual is an epitome of the phylogenetic history of the species (Buckman, 'Mon. Ammon. Inf. Oolite,' Pal. Soc., 1905, p. cc). Similar observations have been made with regard to Brachiopods (Buckman, 'Quart. Journ. Geol. Soc.,' vol. lxiii, 1907, p. 338); primitively the Brachiopod shell is smooth externally, more advanced forms are progressively costate and then spinous. Occasionally species may regress towards a primitive plain form through a costate phase. The progression or elaboration is known as "anagenetic" development, and the retrogression as "catagenetic."

It is regrettable that our present state of knowledge of Chalk Asteroids does not allow us to recognise such definite phylogenetic series as those obtainable in Brachiopods or Ammonites. The plates of an Asteroid are constantly being eaten away and replaced by new calcareous matter, so that the adult plate may differ considerably in character from its young phase. An opportunity for study, however, is afforded by the fact that all the plates are not formed at once. The more distal plates are younger than the proximal plates, and therefore resemble more closely those of the young form. The resemblance is not, however, quite exact, as they are formed later in the life of the individual, and may show consequently characters which have appeared later in the history of the species. Doubtless, if it were not for the paucity of the well-preserved specimens of Cretaceous Asteroids much might be still made out by a comparative study along these lines.

The following paragraphs are only suggestions made in the hope that more material may come to light at a future date. The great majority of starfishes are and have been spinous forms, and I propose to assume that the original ancestor in each group was spinous.

Genera—METOPASTER AND MITRASTER.

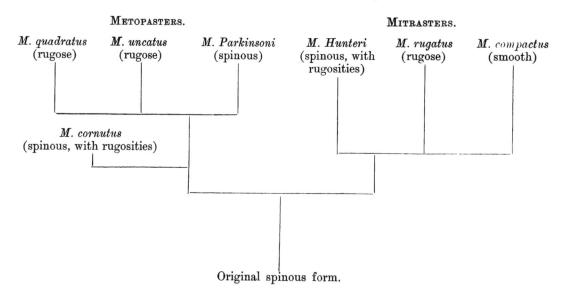

It is convenient in this group to consider the ornament on the raised central area of the marginalia. It will be seen from the above diagram that the spinous form, which was the ancestor of these two groups, early gave rise to two offshoots, one of which includes the *Metopaster* species, the other the *Mitraster* species. Both genera retained evidence of common ancestry by the possession of similar ornament of the specialised "pustulate" type, and by the similar appearance of their ultimate supero-marginalia, which, instead of being smaller than the rest of the superior marginal series, are as large or larger than these. *M. cornutus*, which is the only species found in the lower zones of the chalk (Turonian), shows that early specialisation set in. This form is spinous without rugosities when young (see p. 55, Pl. XIV, fig. 5), but older specimens (see Pl. XXIX, fig. 12) acquire rugosities.

The species of *Metopaster*, *M. quadratus* and *M. uncatus*, have lost all spines from the raised central area of their supero-marginalia, and, instead, possess rugosities. The raised central areas of the infero-marginalia possess neither spines nor rugosities, but are quite smooth. *M. quadratus* has also acquired, as a frequent variation, a primitive type of ultimate supero-marginalia (see p. 98). The type of ornament shows the species to be highly specialised, and this fact, together with its occurrence in the higher zones of the Chalk, affords us an explanation of the remarkable peculiarity of the terminal supero-marginalia on the supposition that it is a catagenetic tendency.

The species of *Mitraster* show even more decided evidence of the three phases —spinous, rugose, smooth. These alterations only occur on the raised central area of the supero-marginalia. The infero-marginalia appear to pass directly from the spinous to the smooth stage without the intervention of a rugose stage.

Genera—Calliderma, Nymphaster, and Pentagonaster.

This group tends to become smooth both in the Senonian and in the Turonian-Cenomanian.

Turonian-Cenomanian forms include *C. Smithiæ*, *C. latum*, *N. Coombii*, *N. oligoplax*, and *N. marginatus*, which are spinous; *N. rugosus*, which is rugose; *N. radiatus*, which is smooth on the older proximal plates, but rugose on the younger distal plates.

Senonian forms include *P. quinqueloba* and *P. lunatus*, which are spinous; *P. obtusus*, which very often possesses marginalia which have lost the majority of the spines and are almost smooth.

Genus—STAURANDERASTER.

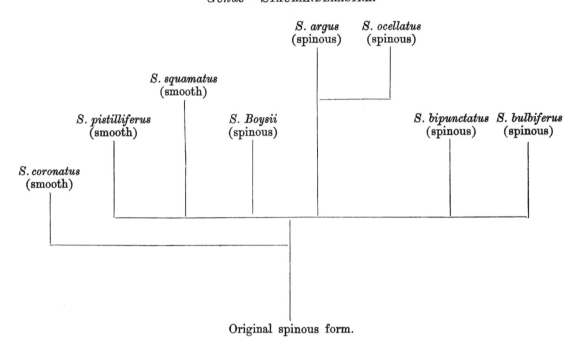

Original spinous form.

The left-hand stem and branches of the diagram above are occupied by the generalised species which have long tapering arms. They show the transition from spinous to smooth forms. The spinous form of these species is *S. Boysii*. In the lower zones of the Chalk one species (*S. coronatus*) appears. This has neither spines nor rugosities on the majority of its marginalia, although a few distal marginalia are rugose. *S. squamatus*, which is almost identical in character with *S. coronatus*, except that it is of smaller size, and *S. pistilliferus* (arms not known) are the smooth forms which characterise the upper or middle zones of the Chalk.

The right-hand stem and branches are occupied by various specialised forms from the upper and middle zones of the Chalk. *S. bulbiferus* shows a specialisation in the bulbiform character of the extremity of its arms, and *S. bipunctatus* in the character of the spines on the ventro-lateralia. *S. ocellatus* and *S. argus* are specialised in the peculiar nature of their armature. All these forms are spinous, their specialisation lying in other directions.

Genus—HADRANDERASTER.

The majority of the plates on this form are spinous, although a few distal plates are smooth, probably indicating a catagenetic tendency in this direction.

Genus—PYCINASTER.

Spines are very feebly developed in this genus. The spine-pits when present are very shallow, and often they are absent altogether. Spine-pits are often visible on the actinal plates after they have disappeared from the abactinal series.

P. crassus possesses rugosities on the distal marginalia.

Genus—ARTHRASTER.

A. cristatus possesses both spines and rugosities; *A. Dixoni* is rugose without spines.

Speculation as to phylogeny in these latter three genera, in view of the state of our knowledge, would be valueless.

GLOSSARY.

The following glossary and diagram (Text-fig. 34) is added to aid the geologist who has but little acquaintance with modern zoological terms.

Abactinal.—Applied to the surface which is uppermost when the starfish walks on its tube feet; the term " dorsal " is used by some authors in the same sense.

Actinal.—Applied to the surface which is undermost when the starfish walks. On this surface are situated the mouth and the ambulacral grooves. The term is used synonymously with " ventral " by some authors.

Adambulacralia or *Adambulacral Plates.*—The ossicles which are adjacent to the ambulacral ossicles. In the order " Phanerozonia," to which the great majority of Chalk Asteroids belong, these ossicles are visible on the actinal surface, bordering the ambulacral groove and hiding the ambulacral ossicles. Adambulacralia may be recognised by their prominent armature of spines.

Adradialia.—Ossicles situated on either side of the radialia (*q. v.*).

Ambulacral.—The ambulacral groove is the groove stretching from the mouth to the extremities of the arm. It is formed by the ambulacral ossicles, which meet in the middle so as to form an arch. The tube feet project through the arch and into the groove.

Centrale.—The most central ossicle on the abactinal surface of the disc. This ossicle, together with five ossicles situated inter-radially and called the "Primary Inter-radialia," are especially prominent in the young form, in which they often occupy almost the whole of the abactinal surface. Generally speaking

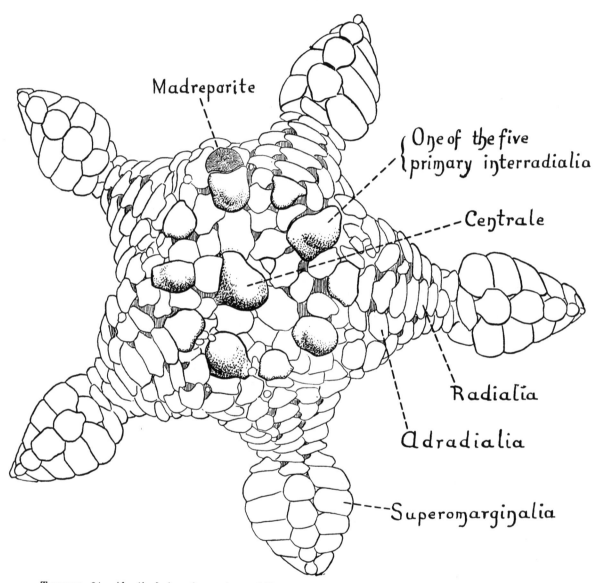

TEXT-FIG. 34.—Abactinal view of a specimen of *Stauranderaster bulbiferus*, natural size, slightly restored from the specimen from Charlton, Kent, registered E. 4344 in the British Museum (Nat. Hist.).

they can be distinguished in the adult form by their larger size, and occasionally they are especially prominent, as in species of *Stauranderaster* (Text-fig. 34) and of the recent genus *Pentaceros*.

Marginalia.—In adult forms of the order "Phanerozonia," which includes

PLATE XXVII.

Linckia sp., n. sp.   (Page 100.)

From the Lower Chalk.

Fig.
1.   Actinal aspect; natural size.   (Coll. Brit. Mus., E. 5055.)
 a. Actinal aspect of portion of arm; magnified 4 diameters.

Ophiura Fitchii, n. sp., *ex Forbes*, MS.   (Page 103.)

From the Flint Gravel.

2.   Actinal aspect, natural size.   (Coll. Norwich Mus.)
 a. Cast of actinal aspect in region of mouth; magnified 4 diameters.
 b. Abactinal aspect; natural size.

Ophiura serrata, *Roemer*.   (Page 102.)

From the Upper Chalk.

3.   Abactinal aspect; natural size.   (Coll. Brit. Mus., E. 5043.)
 a. Abactinal aspect of segment of disc and two arms; magnified 4 diameters.
 b. Abactinal view of two isolated vertebral ossicles of another specimen; magnified 9 diameters.   (Coll. Brit. Mus., E. 5046.)
 c. Side view of the same ossicles; magnified 9 diameters.
 d. Anterior view of the same ossicles; magnified 9 diameters.
 e. Posterior view of the same ossicles; magnified 9 diameters.

Ophiura parvisentum, n. sp.   (Page 103.)

From the Upper Chalk.

4.   Abactinal aspect of type specimen; natural size.   (Coll. Brit. Mus., E. 5052.)
 a. Abactinal aspect of portion of one arm; magnified 6 diameters.

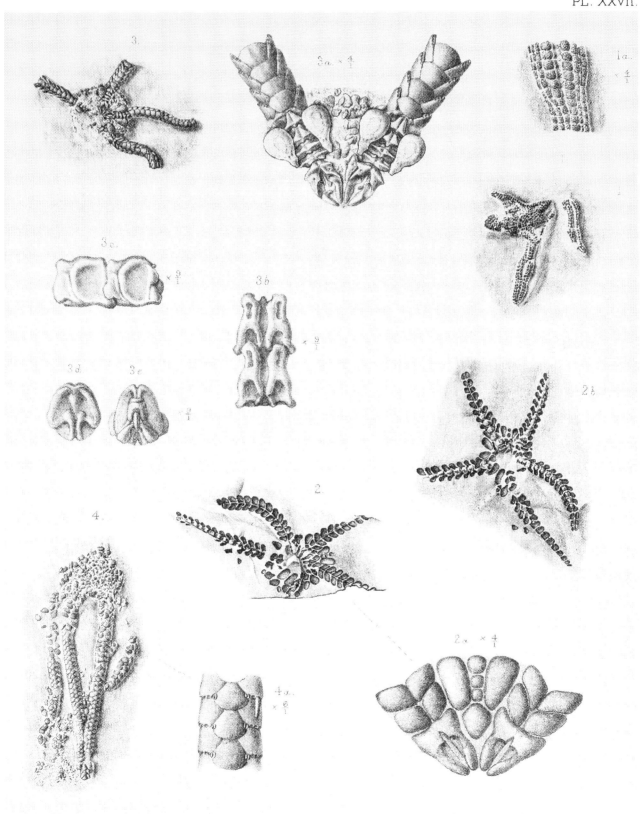

A.H.Searle del. et lith.

Pitcher L^{td} imp.

CRETACEOUS ASTEROIDEA.

PLATE XXVIII.

OPHIOTITANOS TENUIS, n. sp.   (Page 104.)

From the Lower Chalk.

FIG.

1.   Abactinal aspect of type specimen; natural size.   (Coll. Brit. Mus., E. 5056.)
 a. Abactinal aspect of a segment of disc and two arms; magnified 8 diameters (slightly restored).
2.   Actinal aspect of another example; natural size.   (Coll. Brit. Mus., E. 5057.)
 a. Actinal aspect of a segment of disc and two arms; magnified 8 diameters (slightly restored).

OPHIOTITANOS LÆVIS, n. sp.   (Page 105.)

From the Lower Chalk.

3.   Abactinal aspect; natural size.   (Coll. Brit. Mus., E. 5053.)
 a. Abactinal aspect of a segment of disc and two arms; magnified 8 diameters (slightly restored).
4.   Abactinal aspect of another small example.   (Coll. Brit. Mus., E. 5058.)
 a. Abactinal aspect of a segment of disc and one arm; magnified 10 diameters (slightly restored).

OPHIOTITANOS MAGNUS, n. sp.   (Page 106.)

From the Lower Chalk.

5.   Actinal aspect of type specimen; natural size.   (Coll. Brit. Mus., E. 5060.)
 a. Actinal aspect of portion of arm; magnified 4 diameters (slightly restored).

AMPHIURA CRETACEA, n. sp.   (Page 107.)

From the Lower Chalk.

6.   Actinal aspect of type specimen; natural size.   (Coll. Brit. Mus., E. 5059.)
 a. Actinal aspect of a segment of disc and two arms; magnified 10 diameters (slightly restored).

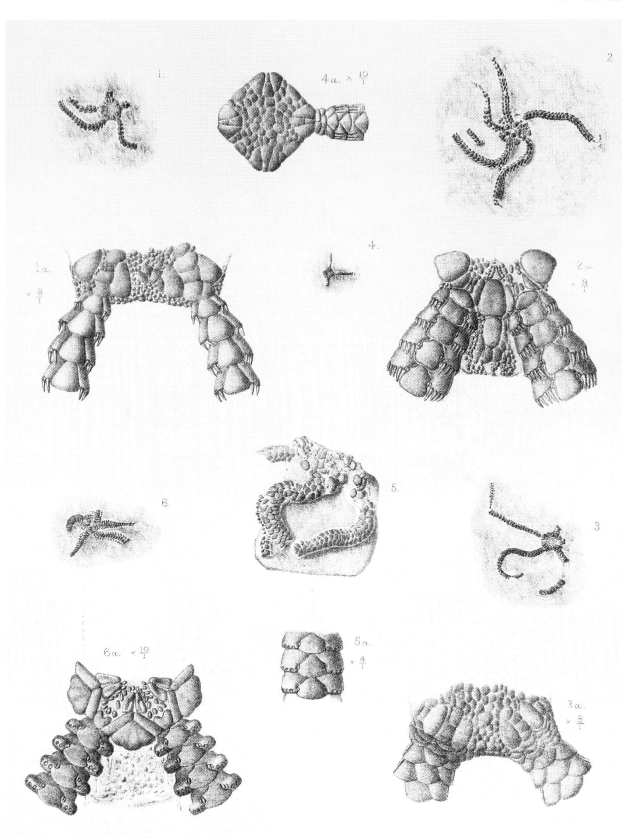

A.H.Searle del. et lith.

Pitcher L^{td} imp.

CRETACEOUS ASTEROIDEA.

PLATE XXIX.

Pycinaster crassus, n. sp. (Page 96.)

From the Upper Chalk.

Pycinaster senonensis, *Valette*, sp. (Page 95.)

From the Upper Chalk.

Nymphaster rugosus, n. sp. (Page 94.)

From the Lower Chalk.

Stauranderaster argus, n. sp. (Page 99.)

From the Upper Chalk.

Arthraster cristatus, n. sp. (Page 93.)

From the Upper Chalk.

Arthraster Dixoni, *Forbes*, sp. (Page 91.)

From the Upper Chalk.

Metopaster cornutus, *Sladen*, sp. (Page 117.)

From the Upper Chalk.

Ophiotitanos magnus, n. sp. (Page 106.)

From the Lower Chalk.

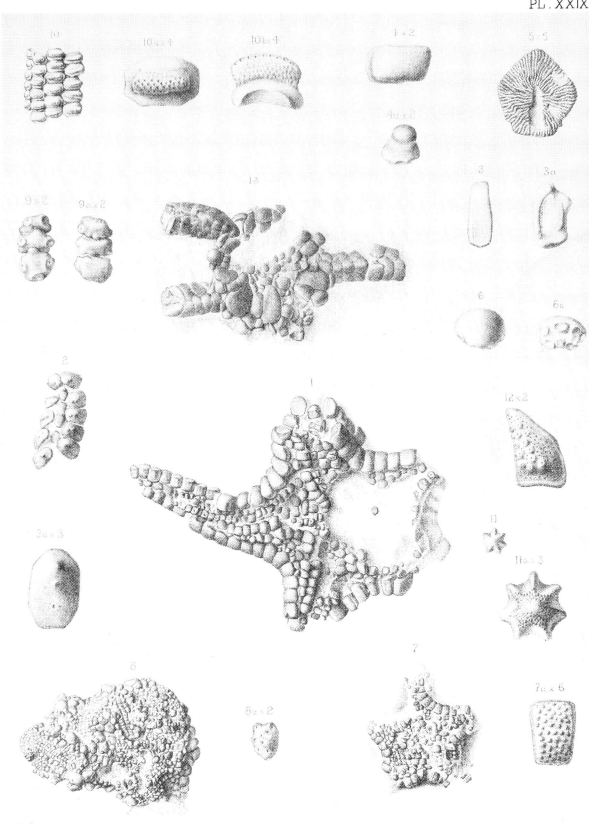

A.H Searle del. et lith.

Pitcher Lᵗᵈ imp.

CRETACEOUS ASTEROIDEA.

Palæontographical Society, 1908.

A MONOGRAPH

ON THE

BRITISH FOSSIL

ECHINODERMATA

FROM

THE CRETACEOUS FORMATIONS.

VOLUME SECOND.
THE ASTEROIDEA AND OPHIUROIDEA.

BY

W. K. SPENCER, B.A., F.G.S.

PART FIFTH.
PAGES 133—138; TITLE-PAGE AND INDEX.

LONDON:
PRINTED FOR THE PALÆONTOGRAPHICAL SOCIETY.
1908.

PRINTED BY ADLARD AND SON, LONDON AND DORKING.

almost all Cretaceous Asteroids, and in quite young forms of the order " Crypto-zonia," the margin of the disc and arms is bordered by specially prominent plates —the " Marginalia." The abactinal series are called " Supero-marginalia," and the actinal series the " Infero-marginalia."

Primary Inter-radialia.—See *Centrale.*

Radialia.—The abactinal series of plates along a major radius are called the radialia.

Radius.—A line drawn from the central point of the disc to an extremity of the arm is called the " *Major radius,*" R. A line drawn from the central point of the disc to a point half-way between two radii is called the " *Minor radius,*" r. This is sometimes called an " *Inter-radius.*"

Spine-pits.—Depressions in a plate for the articulation of spines (see p. 113).

Ventro-lateralia.—The plates on the actinal surface of the Asteroid excluding the infero-marginalia and the adambulacralia. In the inter-radial regions these plates are often rhomboidal. A typical views of an isolated plate of this description is given (Pl. XXIX, fig. 4).

ADDENDA ET CORRIGENDA.

Page 24, line 19, for *Schülze* read *C. F. Schulze.*

Page 26, *Locality, etc.,* for *Upper White Chalk near Norwich* read *Hard Chalk, West Norfolk, the precise locality unknown.*

Page 67, line 11, for *Goniaster compactus* read *Goniaster (Goniodiscus) compactus,* and omit all reference to Forbes, 1848.

Page 69, line 15, for *Stellaster comptoni* read *Goniaster (Stellaster) Comptoni.*

Page 71, line 4, for *Stellaster elegans* read *Goniaster (Stellaster) elegans.*

Page 89, line 8 from end, for *Bourguetiocrinus* read *Bourgueticrinus.*

Page 90, last line, for *Upper Greensand* read *Lower Chalk.*

Page 95, line 3, for *Sladen* read *Forbes.*

Page 95, line 13, for (*p. 89*) read (*p. 89, Pl. XXV, fig. 7*).

Page 101, line 9 from end, the first reference should read OPHIURA, *Lamarck,* 1801. *Systeme des Animaux sans Vertebrés,* p. 350. The date of the reference given is 1816.

Page 102, line 12, for *1841* read *1840.*

Page 103, line 10, add *cor-anguinum* zone, *Northfleet, Kent,* and *Blandford, Dorset*.

Page 103, line 11, add in synonym: Ophiura serrata? *Forbes,* 1843. *Proc. Geol. Soc., vol. iv, p.* 234.

Page 103, line 4 from end, for *parvisentum* read *parvisentis*.

Page 106, line 7, after *E5060* add *and E5061*; under *Locality* insert *Folkestone*.

Page 117, line 12 from end, for *M.* read *Metopaster*.

Page 117, line 7 from end, for *M.* read *Mitraster*.

Page 117, line 2 from end, for *M.* read *Metopaster*.

Page 118, line 3 from top, for *M.* read *Mitraster*.

Page 118, line 7 from top, for *M.* read *Metopaster*.

Page 118, line 9 from top, for *M.* read *Metopaster*.

Page 118, line 12 from top, for *M.* read *Mitraster*.

Page 119, line 3 from end, for *bipunctatus* read *bispinosus*.

Page 120, line 13 from end, for *bipunctatus* read *bispinosus*.

Page 130, table and line 7 from end, for *bipunctatus* read *bispinosus*.

Page 121, lines 2, 4, 6, for *S.* read *Stauranderaster*.

Pl. IV, figs. 2—4, for *Lower* read *Upper*.

Pl. V, fig. 1, for *Tomidaster sulcatus* read *Calliderma Smithiæ* (see p. 123).

Pl. VII, figs. 1 *a* and 2 *a*, for *? Nymphaster Coombii* read *Calliderma Smithiæ* (see p. 122).

Pl. X, fig. 4, for *Metopaster Parkinsoni* read *Metopaster uncatus* (see p. 124).

Pl. XIX, fig. 3, for *Nymphaster Coombii* read *Calliderma Smithiæ* (see p. 122).

Pl. XXI, fig. 2, for *Pentagonaster robustus* read *? a young form of Pycinaster angustatus* (see p. 95).

Pl. XXIV, fig. 1, for *Pentaceros abbreviatus* read *Hadranderaster abbreviatus* (see p. 125).

Pl. XXV, fig. 2, for *Upper Greensand* read *Lower Chalk*.[1]

Pl. XXV, fig. 6, for *Genus? sp.?* (p. 93) read *? Stauranderaster argus* (p. 99).

Pl. XXV, fig. 7, for *Pentaceros? n. sp.* (p. 89) read *Pycinaster angustatus* (pp. 89, 95).

Pl. XXV, fig. 8, for *marginal* read *internal*.

Pl. XXVI, fig. 1, for *Pentaceros punctatus* read *Pycinaster senonensis* (see p. 95).

Pl. XXVI, fig. 4, for *Calliderma mosaicum* read *Pycinaster angustatus* (see p. 95).

Pl. XXVI, fig. 4, for *From the Lower Chalk* read *From the Upper Chalk*.

Pl. XXVII, fig. 3 *b*, for *abactinal* read *actinal* or *adoral*.

Pl. XXVII, fig. 3 *c*, for *side* read *right side*.

---

[1] Mr. H. Woods informs me that recently he has been able to match the matrix in which this fossil is embedded.

Pl. XXVII for *Ophiura parvisentum* read *Ophiura parvisentis*.

Pl. XXVII, fig. 4, before *natural size* insert *slightly less than*.

Pl. XXIX, fig. 12, for *Sladen sp.* read *Sladen*.

On all Plates (except XXVI) for *Calliderma mosaicum* read *C. Smithiæ* (see p. 122).

On all Plates for *Metopaster Bowerbankii, M. Mantelli, M. zonatus*, read *M. Parkinsoni* (see p. 124).

On all Plates for *Metopaster cingulatus* read *M. uncatus* (see p. 124).

On all Plates for *Pentaceros bulbiferus, P. Boysii, P. coronatus, P. bipunctatus, P. squamatus, P. pistilliferus, P. ocellatus, P. argus*, read *corresponding species of Stauranderaster* (see p. 125).

On all Plates for *Pentagonaster megaloplax*, read *P. quinqueloba* (see p. 108).

INDEX.

21

ADLARD AND SON, IMPR., LONDON AND DORKING.